MW00843532

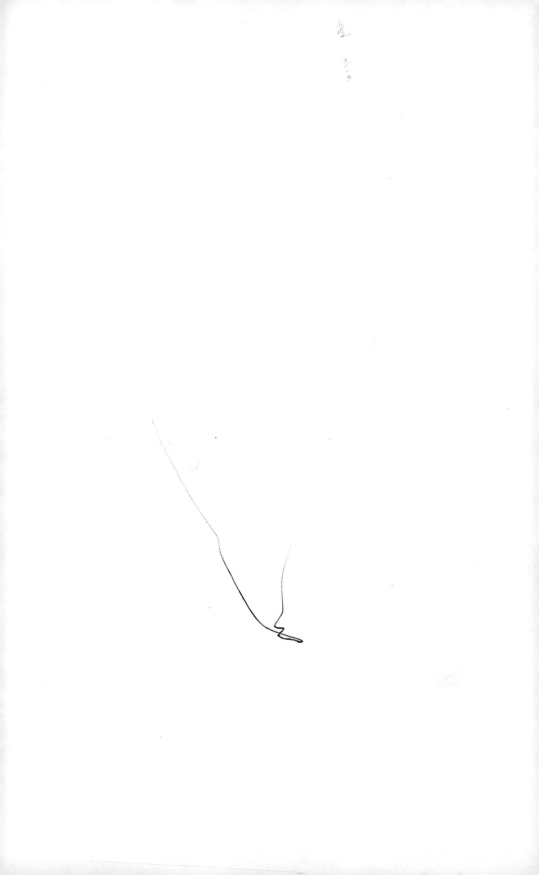

SOILS
in
NATURAL
LANDSCAPES

EARL B. ALEXANDER

CRC Press
Taylor & Francis Group
Boca Raton   London   New York

CRC Press is an imprint of the
Taylor & Francis Group, an **informa** business

CRC Press
Taylor & Francis Group
6000 Broken Sound Parkway NW, Suite 300
Boca Raton, FL 33487-2742

© 2014 by Taylor & Francis Group, LLC
CRC Press is an imprint of Taylor & Francis Group, an Informa business

No claim to original U.S. Government works

Printed on acid-free paper
Version Date: 20130830

International Standard Book Number-13: 978-1-4665-9435-7 (Hardback)

This book contains information obtained from authentic and highly regarded sources. Reasonable efforts have been made to publish reliable data and information, but the author and publisher cannot assume responsibility for the validity of all materials or the consequences of their use. The authors and publishers have attempted to trace the copyright holders of all material reproduced in this publication and apologize to copyright holders if permission to publish in this form has not been obtained. If any copyright material has not been acknowledged please write and let us know so we may rectify in any future reprint.

Except as permitted under U.S. Copyright Law, no part of this book may be reprinted, reproduced, transmitted, or utilized in any form by any electronic, mechanical, or other means, now known or hereafter invented, including photocopying, microfilming, and recording, or in any information storage or retrieval system, without written permission from the publishers.

For permission to photocopy or use material electronically from this work, please access www.copyright. com (http://www.copyright.com/) or contact the Copyright Clearance Center, Inc. (CCC), 222 Rosewood Drive, Danvers, MA 01923, 978-750-8400. CCC is a not-for-profit organization that provides licenses and registration for a variety of users. For organizations that have been granted a photocopy license by the CCC, a separate system of payment has been arranged.

Trademark Notice: Product or corporate names may be trademarks or registered trademarks, and are used only for identification and explanation without intent to infringe.

Library of Congress Cataloging-in-Publication Data

Alexander, Earl B.
  Soils in natural landscapes / author: Earl B. Alexander.
    p. cm.
  Includes bibliographical references and index.
  ISBN 978-1-4665-9435-7 (hardcover : alk. paper)
  1. Soils. 2. Soil ecology. I. Title.

S591.A59814 2014
631.4--dc23   2013031821

Visit the Taylor & Francis Web site at
http://www.taylorandfrancis.com

and the CRC Press Web site at
http://www.crcpress.com

Contents

Preface

Rocks and soils are integral parts of terrestrial ecosystems. They must be considered in any complete investigation of terrestrial ecosystems. Vascular plants, which are the primary producers in terrestrial ecosystems, cannot thrive without roots in soils from which they extract water and nutrients. Biotic diversity is highly dependent on soil diversity, and soils are essential for mitigation of the flow of water from the land. Without soils, flash flooding would be a rule rather than an exception. Soils are essential resources, and all who are involved in land resource management should be able to recognize how soils affect the resources and how management affects soils.

There are many good books on soils from an agricultural perspective, for example, Brady and Weil (2008), Gardiner and Miller (2008), and Singer and Munns (2006), and several books on forest soils. Missing are basic soil books for those who are concerned with natural landscapes and ecosystems. This book was written to fill that niche. It is a thorough introduction to the physics, chemistry, and biology of soils and their roles in local to global systems. Although the emphasis is on soils, an appendix summarizes basic information on the rocks that are most important for describing landscapes and understanding ecosystem functions. Some of the figures in this book can be downloaded in color at the CRC Press website: http://www.crcpress.com/product/isbn/9781466594357.

References

Brady, N.C., and R.R. Weil. 2008. *The Nature and Properties of Soils*. Prentice Hall, Upper Saddle River, NJ.

Gardiner, D.T., and R.W. Miller. 2008. *Soils in Our Environment*. Prentice Hall, Upper Saddle River, NJ.

Singer, M.J., and D.N. Munns. 2006. *Soils, an Introduction*. Prentice Hall, Upper Saddle River, NJ.

About the Author

Earl B. Alexander is a retired pedologist, if you can call writing books (first author of *Serpentine Geoecology of Western North America*, 2007) and professional papers retirement. He has a PhD in soils and has worked for the Soil Conservation Service (now NRCS), the California (now Pacific Southwest) Forest and Range Experiment Station, the Food and Agriculture Organization of the United Nations, the University of Nevada, and three regions of the U.S. Forest Service. He has mapped and investigated soils in Ohio, Nicaragua, Honduras, Costa Rica, Colombia, Nevada, California, Oregon, Alaska, and other states and provinces in western North America. Dr. Alexander has developed interpretations for evaluating soil conditions and expected soil behaviors and made and reviewed recommendations for land management. He taught a Saturday soils field trip at UC Berkeley for two spring semesters. More than 75 of Dr. Alexander's papers have been published in professional journals.

1

SOILS, LANDSCAPES, AND ECOSYSTEMS

Soils and the plants that depend on them are great modifiers of our environment. The climate is very harsh in terrestrial areas lacking soils and plants. Plants that get their water and nutrients from soils are the primary sources of food for animals, other than those in the oceans. Another vital function of soils is the storage of water and slow release of it to springs and streams, rather than allowing it to run off of the land rapidly in enormous floods. "Soil is our most underappreciated, least valued, and yet essential natural resource" (Simpson 2007). Because soils are so essential, it behooves us to become sufficiently familiar with them so that we can be good stewards of the land.

Soils are at the interface of solid earth and atmosphere. They are not claimed exclusively by geologists, whose domain is the solid earth, nor by meteorologists, whose domain is the atmosphere, nor by the biologists, whose domain is the living inhabitants of Earth. Yet concerns of all of these disciplines overlap where there are soils. Rocks disintegrate to become soil, and air and meteoric water (defined in the glossary) occupy much of soils, and plant roots and a great menagerie of microorganisms and small animals occupy soils. The soil and its organisms are constantly exchanging gases with the atmosphere. Thus, soils are within the domains of geology, meteorology, hydrology, and biology, but none of these disciplines include all aspects of soils. The science of soils is a separate discipline, but not independent from the other disciplines.

The general concept of soil, what it is, developed very slowly for many centuries. Then our conception of soil developed more rapidly in the last two centuries, along with the rapid development of chemistry and biology. It may not be possible for those other than specialists in the discipline to obtain a comprehensive knowledge of soils, but all those who work in natural environments will want to have some basic knowledge of soils. We can commence by learning what comprises the field of soils and what the major roles of soils in natural landscapes are.

1.1 Concepts of Soils

Interest in soils dates from early civilizations, when food gatherers became food growers and cultivators. Soils were important to them, because soil is the domain of plant roots, and plants are essential for human sustenance. The Chinese ancestors, and inhabitants of Mesopotamia, the Nile River Delta, and the Indus River Basin all recognized different kinds of soils that warranted different kinds of use and management. In the Americas, the Inca, Maya, and Aztec people were all soil savvy. Written records about soils began to accumulate with the Romans (Winiwarter 2006). Pliny the Elder recognized that different soils had different colors and textures, and that vegetation differences were related to soils (Browne 1944). Farmers learned the importance of organic matter and the cycling of organic waste as manure back to soils in order to maintain their fertility, but the natural sciences did not develop appreciably until alchemists became chemists and mineral and fossil specialists became geologists. By the seventeenth century chemists in northern Europe were analysing and classifying soils (Browne 1944). Geologists began mapping soils before there was an independent discipline of soils, generally with regard to their agricultural potential (Eaton and Beck 1820):

> The present, so far as it has come to our knowledge, is the first attempt yet made in this country to collect and arrange geological facts, with a direct view to the improvement of agriculture.

Soils were considered to be a surficial mantle of weathered rocks and deposits from glaciers, streams, lakes, and wind.

Soil became an independent discipline only after it was widely recognized that soils are unique, and that comprehensive knowledge of soils affirms the interactions of their mineralogy, organic components, chemistry, and physical properties. Fallou (1862) recognized the holistic nature of soils and proposed *Pedologie* as a name for a discipline of soils. About the same time, and publishing slightly later, Dokuchaev in Russia and Hilgard in the United States developed concepts of soils as natural features in landscapes (Tandarich et al.

1988; Jenny 1961). Meanwhile, geologists were developing an agricultural branch of geology. But it was along with the mapping of soils in Russia during the nineteenth century and investigations of Hilgard (1860) that the concept of soils as natural entities (or bodies) was initiated. Independently, Shaler (1891) developed a comprehensive view of soils from the mapping of geology and soils in Kentucky:

> This mass of debris, which at first sight seems a mere rude mingling of unrelated materials, is in truth a well organized part of nature, which has beautifully varied and adjusted its functions with the forces which operate upon it. Although it is the realm of mediation between the inorganic and organic kingdoms, it is by the variety of its functions more nearly akin to the vital than to the lifeless part of the earth.

Unfortunately, soil mapping begun by the U.S. Department of Agriculture in 1899 did not follow the more holistic trends developed independently by Fallou, Dokochaev, Hilgard, and Shaler—at least not for the first one-quarter of the twentieth century.

With the development of soil mapping came the development of soil description. To define a soil map unit, it was not enough to designate the geologic unit and surface soil texture. Early in the twentieth century, soils were commonly described in terms of surface soil, subsoil, and substratum (Figure 1.1A). As the concepts of genetic development of soils evolved and it became evident that the layers of soils are systematically related to one another, soils were described as _profiles_ with a sequence of layers that were designated A, B, and

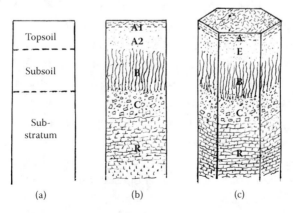

(a) (b) (c)

Figure 1.1 Soil horizon nomenclature evolved along with changing ideas about the depth (a), area (b), or volume (c) necessary to describe a soil.

C horizons (Figure 1.1B) (Bridges 1997). An E horizon was added above a B horizon where iron or aluminum was leached down into the B horizon, generally along with organic matter, causing bleaching of the E horizon. Eventually, more emphasis was placed on the three-dimensional character of soil, and the object of soil description became a *pedon* (Soil Survey Staff 1960) (Figure 1.1C). The description of a pedon is much like that of a soil profile, but with more attention to the lateral variations. Jenny (1980) has suggested adding vegetation and making the unit of description and sampling a *tessera*, an idea that has languished following the passing of Hans Jenny.

1.2 Soil from the Bottom Up and from the Top Down

Disintegrated rocks are the foundations of soils (Chapter 2). Some disintegrated rock, consisting of rock fragments and mineral grains, remains where it is produced by the weathering of solid rock, and some is transported by water, ice, or wind and deposited as alluvium, glacial till, or loess. The cover of unconsolidated residual and transported materials over bedrock is called *regolith* (Greek, *rhegos* + *lithos* = blanket + rock). Soils develop in the upper part of the regolith. Beginning as weathered rock or dirt that is soil parent material, plants send roots into it (Chapter 3), and leaves from plants add organic detritus that is decomposed by small animals and microorganisms to produce humus (Chapters 9 and 10).

Soils differ from rock mainly in that they are unconsolidated and penetrable by plant roots (Figure 1.2), but also in having pores that contain large amounts of air and water (Chapters 3 and 5). A subsoil that has been saturated with water and then drained might have about 40% porosity with 20% water and 20% air in the pore spaces, while the saturated and drained surface soil with about 3% organic matter might have about 60% porosity with 25% water and 35% air. Only 3% organic matter, or humus, greatly increases the porosity over that of subsoil with organic matter < 1% (Figure 1.3). Organic soils that form in wet areas have about 90 to 95% porosity, with pores that are often filled completely by water, excluding air other than that dissolved in the water.

Most of the porosity in soils can be attributed to small pores between sand, silt, and clay particles. Plant roots create tubular pores

Figure 1.2 A moderately deep (65 cm) soil with O, A, B, and C horizons on hard bedrock (R horizon).

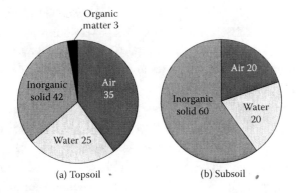

Figure 1.3 Typical proportions (%) of inorganic solids, organic matter, water, and air in (a) surface and (b) subsoils that have been saturated and drained of excess water.

that can be large in surface soils, diminishing in size and abundance down through subsoils. The sand, silt, and clay particles coalesce into aggregates that are commonly small granular aggregates in surface soils and larger blocky aggregates in subsoils (Chapter 3), and these aggregates have pores between them. Small animals, such as earthworms and ants, produce tunnels and chambers. The pores are very important in draining excess water from soils and allowing air to flow more freely through soils.

As water flows though the soil it leaches colloidal particles and dissolved inorganic ions and organic compounds down to accumulate in

Figure 1.4 Earthworms commonly bring soil to the ground surface and deposit it in rounded castoffs.

subsoils. The results of accumulation of organic detritus, conversion of raw organic matter to humus, translocation of colloids and dissolved substances to subsoils, and weathering, slowly form soils with O, A, B, and C horizons (Chapter 3). The uppermost part of the regolith is converted from dirt to soil. Soil is formed by the weathering of parent material at the bottom, addition of organic detritus at the top, and leaching materials downward. The rates of these processes are highly dependent upon soil temperatures and water (Chapters 4 and 5).

Soils are teeming with life, especially topsoils (Chapter 8). A handful of topsoil might contain billions of bacteria, millions of protozoans, thousands of microscopic roundworms, and hundreds of mites and springtails. These organisms are so small that even with their large populations, they generally occupy less than 1% of the pore space in soils. The mass of fungi may be as great as the sum for all other soil organisms. Earthworms, ants, and termites move large amounts of soil (Figure 1.4). Charles Darwin was a lifelong observer of earthworms. He estimated that earthworms brought as much as 1 to 2 mm of soil to the ground surface every year (3 to 6 or 8 inches over 30 years) on degraded and abandoned land, with slower rates at the beginning and again after the worms had developed more favorable habitats for themselves (Darwin 1881).

1.3 Soils, Landscapes, and Ecosystems

Soils are not the same at all locations; they differ from site to site. Many centuries ago, the differences became very important to people

Table 1.1   Comparison of Soil and Plant Organization

Soil	Plant
Primary particle	Cell
Secondary particle (ped)	Tissue
Horizon (layer)	Organ
Pedon	Organism
Polypedon	Community

Although an individual soil is arbitrary, it is a pedon that is classified in *Soil Taxonomy*, just as it is an organism (or its species) that is classified in plant taxonomy.

as they began cultivating soils. Different kinds of soils were given specific names many centuries, or millennia, ago, but the variability and complexity of soils are so great that modern soil classification systems are still evolving (Chapter 6). In the hierarchy of size and complexity of soil features, from primary particles less than a millimeter in diameter to landscape features covering many hectares (Table 1.1), it is the pedon that is classified to identify and name soils.

The Russians and others who covered large areas recognized that soils are different where soil parent materials, climates, and vegetation are different, and that topographic relief and soil drainage are important also. Hans Jenny (1941) popularized this concept in an equation, modified here, where a soil (s) at time (t) is defined by

$$s(t) = f(cl', o', r', p')$$

where cl', o', r', and p' are climatic, organism, and parent material factors. This formula is strictly heuristic, because the soil defining factors are not constant through time, and the properties of a soil are dependent on the path taken to a given state, s(t).

Pedologists who map soils recognise that soils differ across landscapes as they are related to topography, climate, and vegetation differences. They commonly use changes in these landscape features to locate soil boundaries, but not consistently. In the United States, and in other countries, soil surveys have generally been conducted on agricultural lands. The emphasis has been on those soil properties that affect crop production, with less attention given to lithologic substrata, landforms, and vegetation where they are not perceived to be related to crop production. More comprehensive ecological landscape mapping is much less common.

What is landscape? In geomorphology, *landscape* has a more tangible meaning than simply natural scenery as defined in English language dictionaries. It is the surface of the land. The configuration of a landscape is called a landform. Adding *ecological* implies that an *ecological landscape* includes the ecosystem(s) on a landform. Ecological landscapes are the focus of *landscape ecology*, as introduced by Troll (1971) and elaborated by Forman and Godron (1986). Ecological landscape mapping involves lithology, stratigraphy, topography, soils, and vegetation; it warrants much more attention than it has attracted in the past (Chapter 7).

1.4 Global Perspective

Soils are crucial in their influence on the global distributions of water, carbon, and other elements (Chapter 11). Precipitation that falls on the continents, other than Antarctica, and drains to the streams passes through soils, where the concentrations of particulate and dissolved constituents are altered dramatically. Soils remove many humanly undesirable constituents from water and add by-products from soil organic matter decomposition and from weathering in the soils and their parent materials.

Most of the carbon in Earth's crust is tied up in sedimentary rocks, and the deep aphotic zone of the oceans is a very large sink for carbonates. The largest pool of carbon from which it is being continuously cycled to and from other pools is soils. More carbon is stored in soils than in plant communities, the surface euphotic zone of the oceans, or the atmosphere. Gains or losses of carbon from soils are very important in modifying the concentrations of CO_2 and CH_4 in the atmosphere, gases that are very important in regulating heat flow from Earth (Chapter 4). They are greenhouse gases.

Nitrogen, phosphorous, and sulfur are important constituents of soils and plants, and they are important in the global cycling of these elements. Large losses of these elements, especially N, from fertilized crops causes major pollution problems in lakes and streams. Another pollution problem addressed in Chapter 11 is acid rain.

The scope and functions of soils are so great that the soil system of Earth has been considered in relation to *Gaia* (Lovelock 1993;

Van Breeman 1993). Whether or not you consider the Gaia concept appropriate, it is an interesting global perspective.

1.5 Land Use

The quality of life on our Earth depends on the commodities and amenities derived from the land. Social views, or "land ethics" (Leopold 1949), and interactions affect land use, and conversely, land use affects social interactions.

Thousands of years ago people considered humanity to be an integral feature of the environment. As civilization developed, people began to think of land as something to be exploited for humanity's benefit. There seems to be a reversal of this trend among some people, back to thinking of humanity as an integral feature of the environment. It is unlikely that there will be a major reversal of environmental concepts, however, unless a catastrophic event decimates the population, because clothing and feeding a densely populated Earth requires intensive production of food and fiber on much of the land.

The prudent choice of land that can be cultivated and managed intensively without its deterioration depends on the soils and other landscape features. It is most efficient to manage the most productive land, or land that has the lowest risk of deterioration, the most intensively. These two decision criteria, productivity and susceptibility to deterioration, are both dependent on the soils, climate, and management constraints. Therefore, any allocation of intensive land use should take into account the soils. Some civilizations have failed because intensive exploitation of erodible soils resulted in their deterioration to such an extent that the soils were no longer sufficiently productive to support the people of the area (Hillel 1991). Allocating land to a suitable use and managing it appropriately for the maximum production of commodities, or for maximizing amenities, requires knowledge of the soils and the effects of management on the soils and on the ecosystems and landscapes in which the soils occur.

1.6 Land Management

Most people are concerned about soils for their uses, rather than because soils are interesting and have unique flora (plants) and fauna (animals).

Engineers are interested in soils as construction materials and for their support to the foundations of buildings. Hydrologists are interested in the storage and flow of water in soils. Environmental geologists are interested in the stability of soils for predicting mass movement (for example, landslides). Economic geologists are interested in soils where iron, aluminum, chromium, nickel, cobalt, uranium, and other elements are concentrated in quantities suitable for mining. There are many more examples, but we are interested mainly in soils for ecological landscape managers, which is the concern of foresters and rangeland managers, among others.

Forest and rangeland managers are most interested in soil as a medium for plant growth. All terrestrial plants, except bryophytes and epiphytes, are rooted in soils. Typically, one-quarter to one-third of a tree's mass and about one-half of grass plants are in soil. Nearly all of the water plants use and most of the nutrients come from the soil. Thus, anyone who is trying to maintain or increase plant production can profit from a thorough knowledge and understanding of soils.

Forest and rangeland management practices have effects on soils that modify their capacities to support plant growth. Even helicopter logging, which has very little direct effect on soils, has indirect effects through alterations in microclimate and nutrient cycling.

The capacity of a soil to provide societal or ecosystem benefits without being degraded is commonly called soil quality (Chapter 12). It is a highly subjective rating of soils that depends on the management expectations for them. The quality of a soil will depend upon the use allocated to it. A perennially wet soil may have a low quality for growing corn, but a high quality for growing marshland plants and purifying water that flows through the marsh. Soils severely degraded by salinization, compaction, erosion, or other means are in poor condition. Soil condition, or soil health, is an assessment of a soil's present capacity relative to its inherent potential to provide societal or ecosystem benefits.

Questions

1. Soils develop in regolith, but not in all regolith. How do soils differ from regolith that lacks soil?

2. What is passed between soils, streams, and lakes and oceans, and the atmosphere? Water? Rocks? Air?
3. Is it a coincidence that concepts of soils being different with different parent materials, in different topographic settings, and in different climates, were developed in large countries (Russia and the United States) with broad ranges of land-scape features?
4. Place the following evolutionary ideas about soils in actual or presumed historical order:
   a. Plants get essential nutrient elements from soils.
   b. Plants extract most of the water that they use from soils.
   c. Some soils are different from others because they have different parent materials.
   d. Soils in different climates are different.
   e. Soils evolve through time.
   Some of these concepts may have been developed simultaneously.
5. Name three chemical elements that are of primary importance in global cycling.
6. Does the future of our civilization depend on prudent land use? Explain your answer.
7. Match the following disciplines or occupations (A to G) with major reasons for interests in soils (a to h). Some disciplines may have more than one major interest in soils.

A. Economic geologist   a. Suitability for supporting building and roads
B. Environmental geologist   b. Suitability for growing trees
C. (Civil) engineer   c. Sources of valuable minerals and elements
D. Hydrologist   d. Suitability for supporting grazing or foraging animals
E. Forester   e. Suitability as filters for removing pathogens
F. Rangeland manager   f. Slope stability, or mass failure (or landslide) hazards
G. Sanitation engineer   g. Suitability as filters for removing toxic chemicals
                               h. Water storage capacity

Supplemental Reading

Forman, R.T.T., and M. Godron. 1986. *Landscape Ecology*. Wiley, New York.

Hillel, D. 1991. *Out of the Earth: Civilization and Life in the Soil*. Free Press, New York.

Huggett, R.J. 1995. *Geoecology*. Routledge, London.

Kellogg, C.E. 1941. *The Soils That Support Us*. Macmillan, New York.

Leopold, A. 1949. *Sand County Almanac*. Oxford University Press, New York.

McNeill, J.R., and V. Winiwarter. 2006. *Soils and Society: Perspectives from Environmental History*. White Horse Press, Isle of Harris, UK.

Simonson, R.W. 1968. Concept of soils. *Advances in Agronomy* 20: 1–47.

Simpson, M.G. 2007. *Dirt: The Erosion of Civilizations*. University of California Press, Berkeley.

2

SOIL PARENT MATERIAL, WEATHERING, AND PRIMARY PARTICLES

Soils develop from the weathering of rock or in unconsolidated deposits, such as stream deposits or sand dunes, that are called *regolith* (defined in the Glossary). That part of the regolith in which a soil develops is called its *parent material*. It can be disintegrated rock, volcanic ejecta, or organic detritus. Organic soils, which have parent materials of organic detritus, are defined in Chapter 6 and characterized in Chapter 10.

Individual soil particles that are pieces of rock or minerals from parent materials are called primary particles. Aggregates of primary particles that are bound together by clay or organic compounds are called secondary particles. Secondary particles are the subject of Chapter 3.

2.1 Parent Material

The initial composition of a soil is dependent on the composition of its parent rock or parent material. Appendix D presents the different kinds of rocks that are most likely to become soil parent materials. The main chemical elements in soils are those in the igneous rocks that are represented in Table 2.1 by silicic granite to less silicic andesite and basalt and least silicic peridotite. Carbon and nitrogen from the atmosphere are added to soils in organic matter. The compositions of sedimentary rocks vary widely from practically monomineralic quartz (SiO_2) sandstone and limestone ($CaCO_3$) to sandstones and shales that have elemental compositions similar to those of the igneous rocks, although shales generally have more aluminum.

Table 2.1   Chemical Composition of Major Igneous Rocks

Element Oxide	Granite	Andesite	Basalt	Peridotite	Limestone
			g/kg		
SiO_2	725	592	506	419	12
TiO_2	2	11	20	<1	<1
Al_2O_3	142	172	143	7	2
Fe_2O_3	9	45	28	27	—
FeO	10	21	86	51	—
MnO	<1	1	2	1	<1
MgO	4	15	68	432	213
CaO	14	49	105	5	302
Na_2O	33	42	24	<1	1
K_2O	55	29	7	<1	<1
P_2O_5	9	5	2	<1	<1
H_2O	3	8	3	47	—

U.S. Geological Survey (USGS) data for igneous rocks and National Institute of Standards and Technology (NIST) data for dolomitic limestone, which is more than one-half carbonate (CO_3^{2-}), or nearly one-half CO_2. Means of one to three samples are rounded to the nearest g/kg. Geologists report the elemental compositions of rocks as oxides, because the oxides plus water add to 100%, or very nearly 100%; oxygen is the most abundant chemical element in the igneous rocks, and in most other common rocks. (Data from Potts, P.J. et al., *Geochemical Reference Material Compositions*, CRC Press, Boca Raton, FL, 1992.)

The primary particles in soils are the smallest particles that can be separated from each other by vigorous stirring or shaking in water containing a dispersing agent. Primary particles of coarse sand and larger sizes are generally pieces of rock, or rock fragments. Smaller primary particles, less than about ½ or 1 mm, are generally mineral grains rather than rock fragments. The minerals in common rocks are listed in Table 2.2. Soil parent material may be derived from the disintegration or alteration of hard bedrock to produce loose or soft material at the site of the soil, a process called weathering, or it may be glacial till, alluvium, colluvium, dune sand, loess, or lacustrine or marine sediments that were never consolidated materials (Table 2.3). These unconsolidated materials can be characterized by their grain size distributions, particle shapes, and stratification of grains. Fossils in sandstones and shales can be evaluated to differentiate marine from lacustrine sediments. Glacial till and colluvium both have broad particle size ranges, with little or no sorting into strata, or sedimentary layers; that is, they are poorly sorted (geology) or well graded (engineering). Eolian sediments, on the other end of the sorting scale,

Table 2.2 Mineral Abundance in Common Rocks[a]

Mineral	Example Composition	Igneous Rocks			Sedimentary Rocks		
		Granite and Rhyol	Diorite and Andes	Gabbro and Basalt	Sandstone	Shale	Limestone
Feldspars	Aluminum-silicates						
K-Feldspar[b]	$KAlSi_3O_8$	+++	+	-	++	-	-
Plagioclase[c]	$(Na,Ca)Al_3Si_5O_{16}$	+	+++	++	+	-	-
Hornblende	$Ca_2Mg_2Fe_2Al_2Si_7O_{22}(OH)_2$	-	+	+	+	-	-
Pyroxenes	$CaMgFe(SiO_3)_3$	-	-	++	-	-	-
Olivine	$MgFeSiO_4$	-	-	+	-	-	-
Quartz	SiO_2	++	+	-	+++	+	-
Micas	Layer silicates						
Muscovite	$K_2Al_6Si_6O_{20}(OH)_4$	+	-	-	+	+	-
Biotite	$KMgFe_2Al_2Si_6O_{20}(OH)_4$	++	-	-	+	+	-
Clays	$Si_4Al_4O_{10}(OH)_8$	-	-	-	+	+++	-
Calcite[d]	$CaCO_3$	-	-	-	+	+	+++

[a] Abundance scale: +++, abundant; ++, common; +, minor constituent; -, sparse or absent.

[b] Orthoclase is the most common K-feldspar. Microcline is common in pegmatite and sanidine is common in basalt.

[c] Plagioclase is a continuous Na–Ca replacement series that ranges from sodic plagioclase in granite to some calcium in granodiorite, much calcium in gabbro, and calcic plagioclase in anorthosite.

[d] Calcite is the main mineral in limestone and dolomite $(CaMg(CO_3)_2)$ is the main mineral in dolomitic limestone (Table 2.1).

Table 2.3 Transported Soil Parent Materials

Medium of Deposition	Transport Agent	Kind of Deposit	Common Landform
Atmosphere	Wind	Sand, loess	Dune
Stream	Water	Alluvium	Floodplain, terrace
Lake	Water	Lacustrine	Lacustrine plain
Ocean	Waves	Marine	Beach
Glacier	Ice	Till	Moraine
Meltwater	Water	Outwash	Outwash plain, kame
Slope	Gravity	Colluvium	Landslide

have relatively narrow ranges of particle sizes, and thus are well sorted (geology) or poorly graded (engineering). Those aeolian sediments that are predominantly sand form dunes, and those that are predominantly coarse silt are called *loess*.

2.2 Weathering

Exposed bedrock may weather, or disintegrate, from the physical stresses of heating and cooling, wetting and drying, or freezing and thawing. Chemical weathering, however, may be more effective in converting bedrock to parent material in most environments, especially where the bedrock is covered by regolith or soil. Many of the physical stresses that rupture rocks are due to chemical processes. For example, the crystallization of salt in fractures forces pieces of rock apart. And the hydration of biotite in granitic rocks can cause expansion that breaks the rocks into a granular debris of coarse sand to very fine gravel size called *grus*. Biological sources of physical stress, due to root growth in fractures, can be effective in the disintegration of rocks.

Weak acids are major agents of chemical weathering. Rainwater is a weak acid, due to the dissolution of carbon dioxide from the atmosphere:

$$CO_2 + H_2O = HCO_3^- + H^+$$

Plant roots are another source of carbon dioxide. Carbonic acid (H^+, HCO_3^-) and organic acids produced by lichens growing on rocks cause differential alteration and dissolution, releasing solid particles of rock. Other sources of acids are the oxidation of ammonium from organic matter to nitrates and the oxidation of sulfides in marine sedimentary and metamorphic rocks to sulfates that hydrolyse in water to produce nitric and sulfuric acids. The soils in tidal wetlands that are drained can become extremely acidic upon the oxidation of sulfides in them.

Depths of weathering range from shallow in very cold and very dry soils to scores of meters beneath humid tropical soils. Deep weathering commonly alters bedrock to soft masses called *saprolite*. Granitic rocks containing biotite are unique in weathering to grus in warm climates. Bedrock weathering generally advances along joints and fractures. Large blocks of practically unweathered rock between soft patches of weathered rock or grus produced adjacent to joints and fractures are called corestones (Ollier 1984). In granitic rocks with widely spaced joints and lacking fractures between the joints, loss of soil and grus around corestones by erosion leaves massive towers called *tors*. Weathering of limestone by dissolution creates cavernous pockets in the bedrock, and commonly caves at the level of the groundwater table.

The oxidation of iron in minerals hastens their disintegration. Oxidation of manganese may have similar effects, but iron is much more abundant in most of the more common minerals in soil parent materials. The oxidation of sulfides, which are present in many marine sediments, produces an acidic environment that is very effective in weathering.

A weathering sequence for the common silicate minerals is indicated by the molar(+) ratio of alkali plus alkaline earth cations (Na + K + Mg + Ca) divided by silicon ions, minus a small fraction of the molar Al/Si ratio. The predicted mineral sequence, from most to least readily weatherable minerals, is generally

Olivine > Pyroxene = Hornblende > Ca-feldspar > Biotite

> Alkali feldspar > Muscovite

but the relative weatherabilities of different minerals is dependent also on environmental conditions; for example, biotite may be readily converted to vermiculite in an acidic, leaching environment, but otherwise has the resistance to weathering indicated by its position in the sequence. Weathering is related to soil temperature, acidity, and leaching. In humid tropical climates where the alkali and alkaline earth cations and silicon are leached from the soils, the ultimate result is bauxite, which consists of hydrated aluminum oxides; in soils from ultramafic rocks that lack aluminum, the ultimate result is iron oxides, mainly goethite ($FeO(OH)$) and hematite (Fe_2O_3).

Disintegrated rock that accumulates where the rock was weathered is called *residuum*. The parent materials of organic soils are predominantly plant detritus. Organic matter from plant detritus accumulates where prolonged saturation of the soils restricts the movement of gases in them, and the depletion of oxygen retards the decomposition of organic matter.

2.3 Primary Soil Particles

Primary particles are mineral grains and rock fragments. The salient attributes of primary soil particles are size, composition, shape, and surface properties.

2.3.1 Particle Size

Primary soil particles range from very large pieces of rock called boulders to microscopic particles called clay (Table 2.4). All particles with nominal diameters > 2 mm are called rock fragments in the U.S. Department of Agriculture (USDA) system of particle size

Table 2.4 Primary Soil Particle Size Classes

	Nominal Diameter		
	USDA[a]	Wentworth[b]	ASTM[c]
Size Class	mm		
Boulder	>600	>256	>300
Stone	250–600		
Cobble	75–250	64–256	75–300
Pebble[d]	2–75	4–64	4.75–75
Granule		2–4	
Sand	0.05–2.0	0.0625–2.0	0.075–4.75
Silt	0.002–0.05	0.004–0.0625	0.005–0.075
Clay	<0.002	<0.004	<0.005

[a] USDA, U.S. Department of Agriculture.
[b] The Wentworth system is utilized by geologists.
[c] The ASTM system is utilized by engineers.
[d] Pebbles in bulk are called gravel.

classification. The rock fragment designation is not always appropriate, however, because the particles > 2 mm in some soils are concretions or fragments of cemented soil (for example, fractured hardpan), rather than real rock fragments. Particles 2 to 250 mm can be called coarse fragments, which may be a better particle size term than rock fragment, because there is no inference about particle composition in *coarse fragment*. The portion of soil with primary particles < 2 mm is referred to as *fine earth*. Fine earth may include small fragments of rock.

The finer clay particles are colloids. They have large surface areas. Particles < about 0.2 to 1 μm, depending on particle shape and density, do not settle in water unless they are subjected to forces greater than that of gravity.

2.3.2 Particle Size Distribution (Soil Texture)

The particles in soils generally span a range of size classes. The particle size distribution of fine earth is represented by 12 texture classes determined by the proportions of sand, silt, and clay (Figure 2.1).

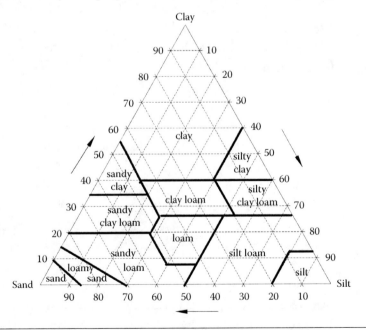

Figure 2.1 A soil textural triangle. The apices represent sand (left), silt (right), and clay (top). Each point on the triangle represents a sample of soil with a certain amount of sand, silt, and clay. Amounts of clay, for example, range from zero at the base to 100% at the top.

Apices of the texture triangle (Figure 2.1) represent 100% sand (left), 100% silt (right), and 100% clay (top). Find the texture class of a soil with 40% sand, 40% silt, and 20% clay. It is a loam, a common surface soil texture. A loam has moderate amounts of sand, silt, and clay, with none predominating. Other texture classes are named as intergrades between sand, silt, clay, and loam. For example, a clay loam has more clay than a loam, but less clay than a clay-textured soil.

The texture of soil with 10 to 35% rock fragments (about 20 to 50% by weight) is modified by gravelly, cobbly, stony, or bouldery, depending on the dominant size of rock fragments. A loam with 35 to 65% gravel or cobbles is a very gravelly loam or very cobbly loam, and with gravel or cobbles > 65%, it is extremely gravelly loam or extremely cobbly loam.

Rock and coarse fragments > 20 mm are generally estimated visually in the field. Finer gravel can easily be sampled for laboratory weighing, or the 2–20 mm gravel can be separated on a 2 mm sieve and weighed in the field with a spring scale. The texture of the fine earth can be estimated by feeling slightly wet soil (near the plastic limit) between the thumb and fingers (Table 2.5). Sandy soil feels gritty. Silty soil feels smooth. Clayey soil holds together well enough that it can be molded into different forms when it is slightly to moderately wet, and when wetter, it is sticky.

2.3.3 Particle Shape

The three-dimensional forms of rock fragments are equant, prolate, oblate, or acicular, depending on whether all axes are about equal (equant), all are different (blade shaped), one axis is shorter than the others (oblate or plate shaped), one axis is longer than the others (prolate), or one axis is much longer than the others (acicular or needle shaped). The edges may be angular, subangular, subrounded, or rounded (Pettijohn 1957). A cube is an equant object with angular edges and a sphere is a rounded equant object.

The shape of a rock fragment is generally indicative of its geologic history. Particles become rounded in stream channels, and commonly more spherical. They also become rounded by wave action on a beach, but not more spherical. Blade-shaped rock fragments are common on

Table 2.5 Soil Texture by Feel

Very Gritty, Usually Noncoherent (Loose When Dry)		Somewhat Gritty			Slightly Gritty to Nongritty						
					Slightly Plastic		Moderately Plastic		Very Plastic		
		Nonplastic	Slightly Plastic	Very Plastic	Rough	Smooth	Rough	Smooth	Rough	Smooth	
Clean	Dirty										
Sand	Loamy sand	Sandy loam	Sand clay loam	Sandy clay	Loam	Silt loam	Clay loam	Silty clay loam	Clay	Silty clay	

Plasticity is the difference between the liquid limit and the plastic limit. The strength of the soil near its plastic limit is assumed to be related to the plasticity. It is estimated by rolling wet soil over a flat surface with the fingers, or between the palm of one hand and the fingers of the other hand if no flat surface is available. Some loams and silt loams may be nonplastic.

beaches. Some rocks, such as shales and slates, break along parallel planes to produce platy rock fragments.

Crystals commonly have distinctive shapes. Most of the clay minerals in soils are platy, but individual clay minerals are too small to see without powerful microscopes.

2.3.4 Particle Composition

The coarse fragments are generally pieces of rock, or rock fragments. Rock fragments may dominate the coarse sand fractions too, but the finer sand fractions are generally individual mineral grains. The finer sand and coarser silt fractions are most commonly dominated by feldspars and quartz. Micas, amphiboles (especially hornblende), and pryoxenes are common, but less abundant, in these fractions.

The clay fractions have a different suite of minerals that are called clay minerals, although quartz may persist in coarse clay. Amphiboles, pyroxenes, olivine, and generally feldspars do not persist in clay fractions. They weather to produce clay minerals such as hydrous mica, or illite, vermiculite, smectite, chlorite, kaolinite, gibbsite, and goethite.

2.4 Clay

Some clay size particles are formed by comminution of larger particles, generally primary minerals that are inherited from the soil parent material. Others, commonly most clay size particles, are formed by the structural alteration of primary minerals or by the precipitation or crystallization of dissolved constituents from soil solution. They are called secondary minerals. Only clay size phyllosilicate minerals—minerals with layer structures—and other secondary minerals in soils, such as gibbsite and goethite, are called *clay minerals*. Quartz, for example, is a primary mineral that is not considered to be a clay mineral even when it is the size of clay.

2.4.1 Clay Minerals

The most abundant clay minerals are aluminosilicates containing aluminum, silicon, magnesium, iron, and oxygen. The basic building blocks are silicon tetrahedra and aluminum, magnesium, or iron

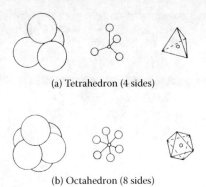

(a) Tetrahedron (4 sides)

(b) Octahedron (8 sides)

Figure 2.2 A silicon tetrahedron and an aluminum octahedron. The large spheres represent oxygen ions that surround the smaller cations.

octahedra (Figure 2.2). Both the tetrahedra and octahedra have central oxygen atoms at their apices and cations between them. There are four oxygen atoms in tetrahedra (four sides) and six oxygen atoms in octahedra (eight sides). Cations from 0.22 to 0.41 times the size of oxygen atoms (for example, Si^{4+}) fit snugly in tetrahedral positions, and those from 0.41 to 0.73 times the size of oxygen atoms (for example, Al^{3+} and Mg^{2+}) fit well in octahedral positions.

Tetrahedra are joined by sharing oxygen atoms between adjacent tetrahedra to form tetrahedral sheets, and octahedra are joined to form octahedral sheets (Figure 2.3). The common aluminosilicate clay minerals are layer silicates (phyllosilicates). Layers are formed by combinations of two or three sheets. A 1:1 layer phyllosilicate, for example, is formed by sharing oxygen atoms in a tetrahedral sheet with

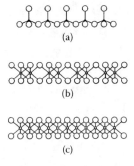

(a)

(b)

(c)

Figure 2.3 Sheets of silicon and aluminum tetrahedra and octahedra. (a) A tetrahedral sheet. (b) A dioctahedral sheet with trivalent cations, usually aluminum, in two-thirds of the octahedral positions. (c) A trioctahedral sheet with divalent cations, commonly Mg^{2+} and Fe^{2+}, in all of the octahedral positions.

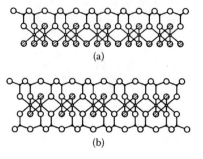

Figure 2.4   Layer silicate clay minerals. (a) A 1:1 layer silicate, such as kaolinite. (b) A 2:1 layer silicate, such as smectite. Notice that the octahedral sheets are dioctahedral. Kaolinite, a dioctahedral aluminosilicate, is the most common 1:1 layer silicate in soils. Serpentine is a 1:1 layer trioctahedral magnesiosilicate. The octahedral sheet in smectites can have a combination of trivalent and divalent cations and not be strictly either trioctahedral or dioctahedral.

oxygen atoms in an octahedral sheet (Figure 2.4). A layer consists of one octahedral and one tetrahedral sheet in kaolinite and halloysite, and one octahedral and two tetrahedral sheets in micas, smectite, vermiculite, and chlorite. Aluminum commonly substitutes for silicon in tetrahedral positions and Fe and Mg substitute for aluminum in octahedral positions. Many other substitutions are possible, but these are the common ones. Substitutions of Al^{+3} or other trivalent cations for Si^{+4} in the tetrahedral layers of 2:1 layer clays create charge imbalances that must be balanced by cations between layers or on the surfaces of the clay minerals. In micas, the usual interlayer cation is potassium (K^+). It fits snugly in the space between six oxygen atoms of one layer and six of the adjacent layer and holds them together closely, preventing the exchange of cations. The interlayer cations in smectite and vermiculite are hydrated and exchangeable cations, generally Mg^{+2}, Ca^{+2}, and Na^+. Smectites absorb water between layers, causing them to swell. Vermiculites have greater tetrahedral sheet charge that prevents them from expanding. Chlorites have a positively charged hydroxyl layer, which is similar to an octahedral sheet, between each set of negatively charged 2:1 layers. The interlayer hydroxyl layer cations of chlorite may be Al^{+3}, Mg^{+2}, Fe^{+3}, Fe^{+2}, Mn^{+2}, or Ni^{+2}, but are generally Mg^{+2} in hydrothermally altered igneous and in metamorphic rocks, and Al^{+3} and Fe^{+3} in chlorites produced by weathering in soils.

A clay mineral that is found only in soils of dry regions is palygorskite. It is a Mg-silicate with a composition similar to that of smectite, but with slices of the 2:1 layer forming chains that are

Table 2.6 Properties of Common Clay Minerals

Mineral Name	Layer[a] Composition tetra:octa	Surface Area m²/g	CEC[b] Constant meq/100 g	CEC[b] Variable meq/100 g
Allophane[c]	0:0	500	<10	40–50
Goethite	0:0		0	<6
Gibbsite	0:1	—	0	<8
Kaolinite	1:1	10–20	1–2	3–6
Halloysite, hydrated	1:1	35–70	5–10	15–30
Hydrous mica or illite	2:1	70–120	15–20	10–15
Vermiculite	2:1	600–800	115–150	<5
Smectite	2:1	600–800	80–120	<5
Chlorite[d]	2:2	70–150	10–30	<20
Palygorskite	Chain	140–190	5–30	<5
Humus[e]	Organic	—	Variable, 150–300	

Note: Most of the common soil clay minerals are layer silicates, except Al- and Fe-oxyhydroxides and allophane.

[a] Ratio of tetrahedral to octahedral sheets in layers.
[b] Cation-exchange capacity (CEC), measured in neutral or alkaline solution. It can be attributed to constant negative charges that are independent of soil pH or to exchange sites that are more negative at higher soil pH.
[c] Allophane has a substantial anion-exchange capacity at low pH. The CEC increases with pH.
[d] Chlorite includes a diverse group of clay minerals with CEC ranges that may differ greatly from those given here.
[e] Soil organic matter, or humus, is added for comparison. Its CEC is dependent on pH, ranging from <100 meq/100 g in very strongly acidic soils to 200 or more in alkaline soils.

joined by oxygen atoms of the silica tetrahedra (Singer 1989). Some Al^{3+} substitutes for Si^{4+}, giving the clay its negative charge (Table 2.6).

Aluminosilicates lacking crystalline structure, called allophane, are common in volcanic materials of humid areas. Allophane is practically amorphous, or has only weakly organized structure. With sufficient structural organization it becomes imogolite.

Iron oxides and oxyhydroxides are common in soils, but generally much less abundant than aluminosilicate clay minerals. The colors of soils with little organic matter are generally determined by these iron minerals. Goethite (α-FeOOH), the most common Fe mineral in soils, is yellowish brown, ferrihydrite ($5Fe_2O_3 \cdot 9H_2O$) is reddish brown, hematite (α-Fe_2O_3) is red, maghemite (γ-Fe_2O_3) is red to brown, and lepidocrocite (γ-FeOOH) is orange (Scheinhost and Schwertmann 1999). Grain or crystal size has a pronounced effect on the colors of some of these minerals (Schwertmann 1993). Magnetite (Fe_3O_4) and other spinel group minerals are common in soils, but they are generally primary minerals rather than secondary minerals.

2.4.2 Clay Properties

Clay is the active inorganic fraction in soils, due to its large surface area, and to ionic imbalances within the clay minerals that attract ionically charged elements and compounds to their surfaces. A gram of montmorillonite, a variety of smectite, has 600–800 m² of surface area (Table 2.6), compared to < 1 m² for silt (Figure 2.5). The surface areas of clay minerals are much greater than the 2 m²/g surface area shown in Figure 2.5, because of platy, rather than equant, shapes, and some clay minerals have internal surfaces.

Clay minerals generally impart negative charges to soils. Two common sources of their charges are (1) substitution of Al^{+3} for Si^{+4} in tetrahedral positions, as in smectite, and (2) broken O– bonds at the edges of aluminosilicate layers, as in kaolinite. Clays hold cations, due to these charges. The sum of these charges is the cation-exchange capacity (CEC). Charges attributed to the substitution of Al^{+3} for Si^{+4} are independent of soil pH, but clay minerals with oxygen atoms exposed to soil solution and able to interact with water molecules in the solution have variable charges that are dependent on the pH of

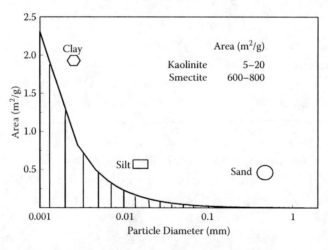

Figure 2.5   Surface areas of equant particles (cubes) from coarse sand to clay size. Layer silicates have more surface area than cubes of equal mass, ranging from somewhat more for kaolinite to much more for smectite.

the soil solution. Minerals with variable charges are minerals with broken O− bonds at the edges of aluminosilicate layers and aluminum and iron oxide minerals. Clay minerals dominated by variable charge sites are negatively charged in high pH soils and become progressively positive as the pH is lowered by the loss of basic cations in highly leached soils. Minerals with substantial variable charge are common in highly weathered and leached tropical soils and in soils developed in weathered tephra in volcanoclastic deposits (Qafoku et al. 2004). The CEC is very important for retaining Ca^{+2}, K^+, and Mg^{+2} that can be exchanged for H^+ and other cations. These basic cations (Ca^{+2}, K^+, and Mg^{+2}) are essential plant nutrients. Soils with net positive charges retain phosphates such that the phosphorous is not available to plants. The kinds and amounts of clay minerals are very important in assessing plant nutrient supplies in soils.

Wet soil with sufficient clay can be molded. The lowest water content at which a soil can be rolled into a 1/8 in. (3 mm) ribbon is called the plastic limit. The water content at which it begins to flow upon agitation is called the liquid limit. The difference between the water contents (%) at the liquid and plastic limits is the plasticity index, which is a soil property important to engineers. Plastic limits range from 20 to 40 for kaolinite, 50 to 100 for smectite, and 130 to 140 for allophane. Liquid limits range from 30 to 110 for kaolinite to 100 to 900% for smectite and 200 to 250 for allophane. Na-saturated smectites have much higher liquid limits than Ca-saturated smectites, but the saturating cation makes little difference for the liquid limit of kaolinite (Mitchell 1976).

Clays like allophane and Na- and Mg-saturated smectites that have high plasticity indices are called *active* clays. They lose strength under stress, a phenomenon called *thixotropy*, and are subject to rapid failure. The "quick clays" of marine sediments that accumulated in shallow glaciated basins, such as those in North America and Northern Europe, that have caused catastrophic mudflows, however, contain more hydrous mica, or illite, and chlorite (Liebling and Kerr 1965). Some of the most well-known quick clays are in Scandinavia and the St. Lawrence River Basin.

The amounts and kinds of clays in soils can be estimated, qualitatively, by feeling soils wetted to their plastic limits to judge soil toughness (called plasticity by pedologists), and much wetter than the plastic limit to judge stickiness. More attention is given to toughness, which is the soil strength at the plastic limit, for estimating the amount of clay; stickiness is more indicative of the kind of clay. Smectites, for example, are much more sticky than kaolinite clay. Although most pedologists refer to soil toughness as plasticity, that is a different attribute of clayey soils, as defined by engineers; plasticity is the amount of water (%) held between the plastic and liquid limits. Nonclayey soils, those with clay < 18 or 20%, approximately, cannot be molded when wet, and therefore do not have plastic indices, or any measurable plasticity, or toughness.

2.4.3 Clay Mineral Distribution

Any of the clay minerals can be inherited from the soil parent material. Micas weather to hydrous micas, or illite, and vermiculite or beidellite (a variety of smectite) by loss of interlayer K^+ and hydration, without destruction of the layer structure. Most other minerals, including montmorillonite (a variety of smectite), are formed by recrystallization of constituents derived from the dissolution of other minerals. The long-term results of weathering are smectites in poorly drained or dry environments where silica and basic cations are concentrated by drainage restriction or evapotransporation. In acid leaching environments, the results are kaolinite and aluminum and iron oxide and oxyhydroxides. Bauxite, which consists of different polymorphs of $Al(OH)_3$, largely gibbsite, formed in warm acid leaching environments, is the main commercial source of aluminum. Allophane forms in the initial stage of the weathering of volcanic glass in humid climates.

2.5 Ion Exchange and the Soil Solution

A very important function of clay minerals and soil organic matter is the retention of basic cations such that they are not readily leached from soils yet are available to plants. The cation-exchange capacity (CEC) is the sum of all cations, both basic and acidic, that can be readily displaced from soil particles and replaced by other cations. The usual basic cations in soils are Ca^{+2}, Mg^+, Na^+, and K^+. The acidic cations, which are H^+ and Al^{3+}, are generally present in soil solutions as hydrated ions.

Table 2.7 The pH Scale

Soil pH Designation	pH Scale	Common Substance
Extremely acidic	0	Battery acid
	1	Sulfuric acid, gastric acid
	2	Lemon juice (pH 2.3)
	3	Vinegar (pH 2.9)
Very strongly acidic	4	Acid rain[a]
Strongly acidic	5	Black coffee
Moderately acidic	6	Urine, uncontaminated rain[b]
Neutral	7	Pure (distilled) water
Moderately alkaline	8	Baking soda (pH 8.4)
Strongly alkaline	9	
Very strongly alkaline	10	Milk of magnesia (pH 10.5)
	11	
	12	Household ammonia (pH 11.9)
	13	Lye
	14	Caustic soda

[a] The pH of acid rain can be down two or more pH units from that of uncontaminated rain.

[b] Rain equilibrated with the current CO_2 concentration of the atmosphere is at pH 5.6, down from about 5.7 before the industrial age.

Some acidic soils with large proportions of aluminum or iron oxides, or hydroxides, have substantial anion-exchange capacities. More important for plants, however, the aluminum and iron compounds bind nutrients that are present as oxyanions (for example, $H_2PO_4^-$ and HMO_4^-) and make them unavailable to plants.

The sources of water in soils are from precipitation, which is normally moderately acidic (Table 2.7). Many soils receive water drained overland or through the regolith from higher landscape positions. In the initial stages of weathering in rudimentary soils, the soil solution will contain bicarbonates of soluble cations derived from the weatherable minerals in the parent materials. If the basic cations are leached from soils more rapidly than they are replaced, decomposition of organic matter in plant detritus acidifies the soil solution and lowers the soil pH. This is why well-drained soils generally become more acidic as they age. The proportions of alkaline earth cations, Mg and Ca, in the soil solution and on the cation-exchange complex decline less rapidly than the proportions of the alkali cations, Na and K, because more of the Ca and Mg are cycled through the vegetation and returned to the soil when the plant detritus is decomposed. At about pH < 4.8, Al becomes mobilized and is toxic to many soil organisms.

The proportion (%) of exchangeable cations that are basic, referred to as base saturation,

$$BS = 100 \text{ (sum of basic cation)/CEC}$$

is a crude index of leaching intensity. In leaching environments, which are those environments that are not arid, or in closed drainages where cations accumulate, the base saturation declines over millennia. The intensity of leaching is dependent upon the amount and composition of water that infiltrates into a soil, or drains to it from upslope, and the amount and rate of drainage through it. In a readily weathered soil parent material, such as limestone consisting of Ca-carbonate and lacking aluminum, cations that are lost by leaching are replenished from weathering parent material rapidly enough to maintain high base saturation and remain nonacidic until the parent material is reduced to mainly impurities from the limestone and aeolian dust. The base saturation is generally based on the amount of exchangeable cations (CEC) at pH 7.0 or 8.2. In soils with substantial variable charge, the CEC measured at pH 7.0 will be much lower and the base saturation much greater than at pH 8.2. Below pH 4.8 the mobilization of Al renders base saturation a meaningless concept.

In well-drained soils where there is enough water to leach Si and Ca to subsoils, but not entirely from the soils, these elements accumulate in calcite or silica concentrations or layers (Chapter 6). In poorly drained soils with sparse vegetative cover and minimal organic matter, Na-carbonates and Ca-sulfates can accumulate. In poorly drained soils with more vegetative cover and organic matter, the soils are more acidic and Fe can be reduced and sulfides can be produced. If soils containing sulfides are drained, the sulfides can be oxidized to produce acid sulfate soils. Acidic soils are commonly produced when tidal flats are drained (Fanning and Fanning 1989).

Note on Units of Measure

The SI units (Le Système International d'Unités) of CEC are mmol+/kg. The CEC has been reported as meq/100 g in twentieth-century literature. Now it is commonly reported as cmol+/kg, rather than in proper SI units, in order to avoid moving the decimal point one space to the right; that is, 1.0 meq/100 g = 1.0 cmol+/kg = 10 mmol+/kg, where meq is milliequivalents, cmol is centimoles, and mmol is millimoles. The + sign indicates moles of positive charge, which are the moles of a cation multiplied by the charge of the cation.

Problems

2.1. Find classes of soil texture from Figure 2.1.
   a. What is the texture of a soil with 60% sand and 15% clay?
   b. What is the texture of a soil with 25% sand and 15% clay?
2.2. Does a sandy loam with 12% smectite or a clay loam with 38% kaolinite have more surface area, assuming that most of the surface area is on the clay minerals?

Solution: From Table 2.6, assume that the smectite and kaolinite have surface areas of 700 and 15 m²/g. The soil areas are
   • Sandy loam with smectite: 12 g/100 g × 700 m²/g = 84 m²/g of soil
   • Clay loam with kaolinite: 38 g/100 g × 15 m²/g = 5.7 m²/g of soil
The sandy sand with smectite has much more surface area than the clay loam with kaolinite.

2.3. What is the CEC of a soil with 2% organic matter (CEC = 150 meq/100 g) and 28% clay that is half smectite and half kaolinite, 14% of each clay mineral.

Solution: From Table 2.6, assume that smectite and kaolinite have CECs of 100 and 5 meq/100 g. Then, CEC (soil) = 2 g/100 g × 150 meq/100 g + 14 g/100 g × 100 meq/100 g + 14 meq/100 g × 5 meq/100 g = (3 + 14 + 0.7) meq/100 g = 17.7 meq/100 g.

The standard international (SI) unit is 177 mmol+/kg, but 17.7 cmol+/kg is more common in the current (post-milli-equivalent) soil literature.

2.4. What is the base saturation of a soil with 4.5 meq/100 g Ca^{2+}, 3.0 meq/100 g Mg^{2+}, 0.6 meq/100 g Na^+, 0.3 meq/100 g K^+, and 6.3 meq/100 g exchange acidity?

Solution: (4.5 + 3.0 + 0.6 + 0.3 meq/100 g)/(4.5 + 3.0 + 0.6 + 0.3 + 6.3 meq/100 g) = 8.4/14.7 = 0.571 = 57%, rounded to the nearest percent, because the measurement does not warrant greater precision.

Questions

1. Why do shales generally contain more aluminum than do sandstones? Hint: Shales contain more clays, and sandstones contain more quartz and feldspars.
2. Which of the following reactions are hydration and which are hydrolysis?
   a. $HOH + NH_3 = NH_4^+ + OH^-$
   b. $2HOH + NH_4OH = NO_3^- + H^+ + 4H_2$
   c. $3HOH + Al^{3+} = Al(OH)_3 + 3H^+$
3. Stream gravels generally become rounded and somewhat spherical. Why are rounded ocean beach gravels commonly more blade shaped, than spherical?
4. Smectites are common in neutral to alkaline soils, but not in nonsaline acidic soils. Why not?
5. Are smectites or kaolinite more likely to be the dominant clay minerals in subsoils that are grey and highly plastic? Is goethite more likely to be a common mineral in subsoils that are grey and slightly plastic or in subsoil that is yellowish brown?
6. Which clay minerals are most likely to be found in acidic soils?

Supplemental Reading

Filep, G. 1999. *Soil Chemistry, Processes and Constituents*. Akadémiai Kiadó, Budapest.

Jackson, K.C. 1970. *Textbook of Lithology*. McGraw-Hill Book, New York. (One of the many good books on rocks, or petrography and petrology)

Klein, C., and B. Dutrow. 2007. *Mineral Science*. Wiley, New York. (An excellent introduction to the intricacy of minerals, and also a great reference)

Mitchell, J. 1976. *Fundamental of Soil Behavior*. Wiley, New York. (A good book on soils from an engineering perspective—considerable information on rheological properties of soils)

Ollier, C. 1984. *Weathering*. Longman, Essex, England. (A good book on weathering in Earth's crust)

Schulze, D.G. 1989. An introduction to soil mineralogy. In J.B. Dixon and S.B. Weed (eds.), *Minerals in the Soil Environment*, pp. 1–34. Soil Science Society of America, Madison, WI. (A brief review of the basic principles of soil mineralogy)

Sposito, G. 1989. *The Chemistry of Soils*. Oxford University Press, New York. (An excellent introductory soil chemistry book)

Soil Survey Division Staff. 1993. *Soil Survey Manual*. Soil Conservation Service, USDA, Washington, DC. (An authoritative text on the description and mapping of soils)

3

SOIL ARCHITECTURE— STRUCTURAL UNITS AND HORIZONS

Soils are not homogeneous. They are heterogeneous on any scale. Soils generally differ laterally over distances of meters, vertically over distances of centimeters or millimeters, and microscopically over shorter distances. Most soils grade laterally in gradual transitions to other kinds of soils. Small units of 1 to 10 m^2 that are relatively uniform laterally are chosen to represent soils. These units have been called pedons, as shown in Figure 1.1. Within pedons most of the differences are vertical. Different layers are called horizons, because they are generally parallel to the ground surface that represents the horizon between land and sky.

Macroscopic and rheologic properties of soils and the features utilized to describe soils are the focus of this chapter. Soils are described with some terminology from sedimentary petrology and soil mechanics and some terminology that is unique to the discipline of soils.

3.1 Soil Structure

As soils evolve in the detritus of weathered rock or in unconsolidated sediments, cohesive forces develop between primary particles. These forces are attributable to clay, iron oxyhydroxides, organic matter, and other soil materials with large surface areas. They are the forces that bind primary soil particles together.

If there is no cohesion between primary particles, a soil is referred to as structureless. If cohesion is evident and is uniform in all directions, a soil is *massive*. Commonly, cohesion is not uniform, so that soil parts (splits) along planes of weaker cohesion. Aggregates of primary soil particles formed by parting along natural planes of weakness are

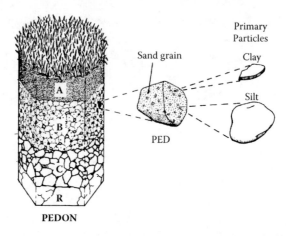

Primary
Particles

Sand grain   Clay

Silt

PED

PEDON

Figure 3.1   A sketch of a pedon with A, B, C, and R horizons. The differences in scale are about 100 from a pedon to peds and greater than 1000 from peds to primary particles. Large sand grains are visible within peds without magnification, but individual very fine sand and silt particles can be distinguished only with magnification.

called *peds* (Figure 3.1). Artificial aggregates produced by shearing or breaking through natural aggregates, or peds, are fragments of soil that are called *clods*—at least that is the name learned on a farm.

The size, shape, and arrangement of secondary soil particles, or peds, constitutes *soil structure* (Soil Survey Staff 1993). Soils lacking peds are considered to be *structureless* (either single grain or massive). Soil structure is generally described by the relative distinctness, size, and shape of peds (Table 3.1). Some peds are coated with clay or other material that may be described as ped surface features. Chemical and mineralogical segregations, such as mottles and concretions that may

Table 3.1   Types of Peds and Their Size Classes

		Type of Ped		
	Platy	Prismatic and Columnar	Blocky (Angular or Subangular)	Granular
	Thickness	Nominal Diameter		
Size Class	mm	mm		
Very fine	<1	<10	<5	<1
Fine	1–2 mm	10–20	5–10	1–2
Medium	2–5 mm	20–50	10–20	2–5
Coarse	5–10 mm	50–100	20–50	5–10
Very coarse	>10 mm	>100	>50	>10

The distinctness classes are weak, moderate, and strong.

Figure 3.2 Examples of soil structure. (A) Granular A horizon. (B) Angular and subangular blocky structure. (C) Slickensides. (D) Platy A horizon. (E) Prismatic Bt horizon. (F) Columnar Bt horizon. (The photographs are cut from those in Soil Survey Division Staff, *Soil Survey Manual*, USDA Handbook 18, U.S. Government Printing Office, Washington, DC, 1993, and Soil Survey Staff, *Soil Taxonomy: A Basic System for Making and Interpreting Soil Surveys*, Agriculture Handbook 436, USDA, Washington, DC, 1999.)

be considered aspects of structure, are commonly described along with soil color or special features.

The most common types of soil structure, blocky peds in subsoils and granular in surface horizons, are illustrated in Figure 3.2, along with other kinds of structure. Granular and blocky peds are equidimensional, as a sphere or a cube, but the peds are multifacial. Prismatic and columnar peds are elongated in the vertical direction. Columnar peds have rounded tops, whereas prismatic peds do not. Platy peds are short in one direction, which is generally vertical; they are flattened horizontally. Granular peds differ from other types of peds in that their surfaces are unrelated to the surfaces of adjacent peds. Blocky peds may have similar shapes, but they differ from granular peds in having surfaces that are accommodated; that is, the faces of a ped are complementary to the surfaces of adjacent peds. Accommodation means that a ped surface is the mirror image of an adjacent ped surface, as though one were a mold and the other a cast. Blocky peds are generally larger than granular peds, but that is not a differentiating criterion.

Granular structure is generally formed by soil organic matter, or humus, binding or sticking the primary soil particles together into aggregates. The more distinct the peds, the stronger the structure. Clay and iron oxyhydroxides are other important agents of cohesion in soils. Soils lacking these agents of cohesion are generally structureless. Soils without pedal structure are called single grained if the primary particles are not held together by cohesion or cementation or massive if they are held together by a binding agent.

Pedal structure other than granular is formed by volume changes that cause the soil to part along planes of weakness to form accommodated peds. The main causes of the volume changes are shrink-swell and freeze-thaw cycles. Soils with greater swelling clay (especially smectite) content commonly have stronger structure. If the stresses accompanying volume change are predominantly horizontal, a prismatic or columnar structure develops. If they are vertical, a platy structure develops. A platy structure commonly develops in fine layers of alluvium that are stratified due to differential sedimentation. Compacted soils commonly develop platy structure, particularly where the compaction was caused by vertical stress such as loading with heavy equipment.

In cracking-clay soils that shrink and swell upon drying and wetting, the movement of large blocks of soil across one another produces smooth surfaces that are called *slickensides* (Figure 3.2). Slickensides are a diagnostic property of cracking-clay soils, or Vertisols (Chapter 6).

Excess Na in alkaline soils causes dispersion of the clays, which can cause deterioration of soil structure and greatly reduce soil permeability. Sodic soils, those with excessive amounts of exchangeable Na, are those with a layer (horizon) containing exchangeable sodium percentages (ESPs) > 15% or sodium adsorption ratios (SARs) > 12; either result indicates a natric horizon and a sodic soil (Soil Survey Staff 1993).

$$ESP = \text{exchangeable } Na^+/\text{sum of exchangeable cations}$$

$$SAR = Na^+/((Ca^{2+} + Mg^{2+})/2)^{0.5}$$

where the cation concentrations are all molar charges (mol_c or mol+).

3.2 Soil Pores and Soil Density

The spaces between solid soil particles are called pores (pedology) or voids (engineering). They are filled by water, air, and plant roots. Small animals and microorganisms generally occupy less than 1% of the pore space.

Pore attributes are size, shape, orientation, continuity, and tortuosity. Many of these attributes are directly related to soil structure (Brewer 1964; Stoops 2003). For example, fine, granular structure implies that there are many very fine and fine, irregular, continuous pores between aggregates; these pores are sometimes called compound packing voids. Spaces between other kinds of peds are planar (but not entirely flat) or irregular, closing when the soil expands. Pore spaces within peds are of many kinds: (1) packing voids between primary particles, (2) irregular voids that are larger than packing voids, sometimes called vughs, (3) tubular pores formed by plant roots and burrowing animals, (4) insect-created chambers along tubes, and (5) closed spherical, or vesicular, pores formed by expanding gas, generally in surface soil.

The sum of all pore space in a soil is its total porosity (P). It is the volume of all pores, or voids (Vv), per volume of soil (Vt):

$$P = Vv/Vt \qquad (3.1)$$

or $\qquad P = Vv/(Vs + Vv)$

where Vs is the volume of solid soil particles.

The conductivity of air and water through saturated soil is more dependent on macroporosity than on total porosity. Total porosity, however, is much easier to determine than macroporosity, because it can be computed from soil particle and bulk densities.

$$P = (1 - Db/Dp) \qquad (3.2)$$

where the bulk density (Db) is the weight of dry soil per volume of moist soil and the particle density (Dp) is the weight per volume of solid particles. Because the most abundant minerals in soils have particle densities near that of quartz (2.65 Mg/m^3), it is usually satisfactory to substitute 2.65 for Dp in Equation 3.2 to compute the bulk densities of inorganic soils. Iron compounds with much higher particle densities are common, but generally not abundant in soils. Fresh pumice with many vesicular pores has densities < 1.0, but the vesicular pores are quickly opened by weathering in soils to give particle densities nearer to 2 Mg/m^3. The particle density of soil organic matter is about 1.4 Mg/m^3. Thus, the mean particle densities of organic soils will range from 1.4 Mg/m^3 for those with 100% organic matter to over 2 Mg/m^3 for those with substantial mineral matter. Soil porosities are commonly about 0.5 to 0.7 (50 to 70%) in uncultivated surface soils, generally increasing with greater organic matter content, and about 0.4 to 0.6 in subsoils. The corresponding bulk densities are about 1.3 to 0.7 Mg/m^3 in surface soils and 1.6 to 1.0 Mg/m^3 in subsoils. Compacted soils with very little organic matter have bulk densities up to about 1.8 to 2.1 Mg/m^3. Organic soils with no inorganic, or mineral, content have porosities > 0.7 (70%). Those with high organic fiber contents have bulk densities < 0.1 Mg/m^3, and those with little fiber have bulk densities of about 0.1 Mg/m^3.

The relationship of soil bulk density to texture, in the absence of organic matter, is shown in Figure 3.3. Sandy soils have the highest bulk densities. The influence of soil organic matter on bulk density is

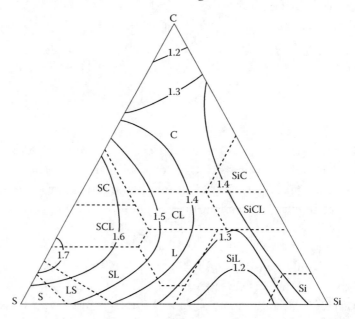

Figure 3.3   Soil bulk densities (Mg/m^3) in relation to texture as transcribed from Rawls (1983) and published in Alexander and Poff (1985). Rawls estimated the bulk densities for soils without organic matter by assuming that each percentage (g/100 g) of organic matter reduced the bulk densities by 0.224 Mg/m^3.

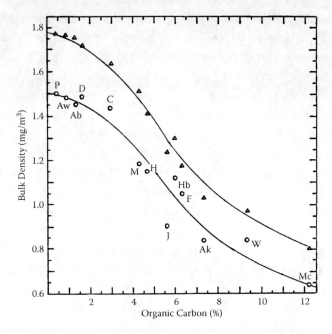

Figure 3.4 Influence of soil organic matter and laboratory compaction on the bulk densities of loamy surface soils. ○ = uncompacted soils; Δ = compacted soils. (Based on data from Howard, R.P. et al., *Soil Science Society of America Journal*, 45, 231–236, 1981. Copied from Alexander, E.B., and R. Poff, *Soil Disturbance and Compaction in Wildland Soils*, Earth Resources Monograph 8, USDA, Forest Service, Pacific Southwest Region, Vallejo, CA, 1985.)

shown in Figure 3.4. Laboratory compaction increased the bulk densities of all of the soils, regardless of organic matter content (Howard et al. 1981).

3.2.1 Note on Units of Measure

The English units for density were replaced first by g/cm^3 in the metric system, and then by Système International (SI) d'Unités of Mg/m^3. Because g/cm^3 is equal to Mg/m^3, the numbers are unchanged.

3.3 Soil Strength

Soil strength is very important to engineers because it is a measure of loads that a soil will support, and it is an indicator of susceptibility to mass failures. It is important to ecologists because great soil strength limits the penetration of roots and of animals that are larger than the pores in soils.

Soil strength is generally dependent on cohesion and the friction between particles in the soils. It is greater in dry soils than in wet soils, because water reduces the cohesion and acts as a lubricant. Cementation that is common in relatively dry areas where silica or calcium carbonates accumulate in subsoils, rather than being leached from them, greatly increases soil strength. Some cemented soil materials are as hard as rocks.

Soils with large amounts of clay, particularly clays like smectites with large surface areas, have much greater strengths when dry than soils that lack clay. The strengths of clayey soils are greatly diminished when they are wet, and clays with smectites are highly susceptible to mass failure and landslides where they are present on slopes. Smectite clays shrink when they dry and expand again when they become wet. The drying and rewetting of smectite clays causes differential movement within soils and the formation of slickensides (Figure 3.2), where movement occurs along inclined (neither horizontal nor vertical) planes. Ground surface expression in the form of low mounds formed by extreme soil expansion and contraction accompanying shrink and swell has been call gilgai relief. Large cracks open to the ground surface when the expanding clay soils dry. These soils have been called cracking-clay soils; they are Vertisols (Chapter 6).

3.4 Ecosystem Effects of Soil Structure

The movements of air, water, and roots in soils are greatly affected by soil structure. Therefore, it is very important to consider soil structure in assessing the hydrology and productivity of soils. Soil bulk densities and total porosities are convenient indicators of soil physical conditions. Plant root penetration decreases as bulk density increases, because soil strength, or resistance to root penetration, is related to bulk density. Few roots will penetrate clayey soils with bulk densities > 1.4 or sandy soils with bulk densities > 1.75 Mg/m^3 (Alexander and Poff 1985). Soil structure is particularly important in dense soils, because many roots may grow between peds where the peds are too dense for root penetration.

Soil porosity is an indicator of permeability and aeration, but not always a good one. One large pore will allow much more water to pass through a saturated soil, or much more air to pass through an unsaturated soil, than many small pores that collectively have a cross-sectional area equivalent to that of the large pore. Pore orientation, tortuosity, and continuity are also very important for the conductivity of water and air.

Porosities between 0.5 and 0.65 are generally quite favorable for plants. Most surface soils are in or near this range of porosities under natural conditions. Subsoils with little organic matter can have lower porosities and great enough strength to inhibit root penetration.

Soil compaction is a major problem in cultivated soils. It is unlikely to be a problem in soils of natural landscapes, however, other than along large animal trails. Problems might be encountered in reclamation projects that involve heavy equipment. The compaction hazard is greatest in moist soils, rather than in wet or dry soils. Water is essentially noncompactable, but puddling can occur in fine-textured wet soils (Chapter 12). Puddling is a process by which soil structure is destroyed by shear forces, leaving the soil without large and continuous pores to conduct water and air effectively.

3.5 Soil Horizons and Pedons

Soils form in material weathered from bedrock or in accumulations of fluvial, glacial, colluvial, or aeolian deposits. That material is called *regolith* until it is occupied by organisms and soil development begins. Soil development is recognized by the differentiation of layers called *horizons* parallel to the ground surface. These horizons are distinguished by color, texture, structure, and chemical properties. For example, a soil may have a black, sandy loam, granular surface horizon; a very acidic, grey subsurface horizon; and a reddish brown, clay loam, prismatic subsoil. Whereas the vertical extent of a soil is relatively well defined by root occupation and development of soil horizons, the lateral extent of a soil is commonly arbitrary. Nevertheless, the concept of an individual soil with finite lateral and vertical dimensions is important in soil description and classification.

3.5.1 Soil Horizons

The main processes of soil development are the addition of plant detritus and its transformation to humus; weathering of the regolith and inorganic soil materials; leaching of dissolved constituents, and in many soils, humus and clay, from upper horizons; and accumulation of clay, humus, and inorganic substances in lower horizons. Horizon nomenclature reflects these processes. An aboveground accumulation of plant detritus is called an O horizon; a surface soil accumulation of humus is called an A horizon; a leached and commonly bleached subsurface layer is called an E horizon; a subsurface layer of advanced weathering or clay accumulation is called a B horizon; relatively unweathered substratum is called C horizon; and hard bedrock is called an R layer (Figure 3.1). These *master* horizons are designated by uppercase letters and are commonly followed by lowercase letters that indicate the horizon properties or soil development processes more specifically (Table 3.2). The O and B master horizons are always

Table 3.2   Soil Horizon Suffix Symbols and Applicable Master Horizon

Suffix Symbol	Master Horizons[a]	Explanatory Caption
a	O	Highly decomposed organic material
b	A, E, B	Buried genetic horizon
c	A, E, B, C	Concretions or nodules
d	B, C	Physical root restriction (e.g., dense till and plow pans)
e	O	Moderately decomposed organic material
f	A, B, C	Frozen soil
g	A, E, B, C	Strong gleying
h	B	Illuvial accumulation of organic matter
i	O	Fresh or slightly decomposed organic material
k	B, C	Accumulation of carbonates, commonly $CaCO_3$
m	B, C	Cementation or induration
n	B, C	Accumulation of sodium
o	B	Residual accumulation of sequioxides
p	O, A	Mixing by tillage or other disturbance
q	B, C	Accumulation of silica
r	C	Bedrock weakly consolidated or weathered soft
s	B	Illuvial accumulation of sesquioxides
ss	A, B, C	Presence of slickensides
t	B	Illuvial accumulation of silicate clay
v	B, C	Plinthite—soft iron concentrations that harden upon repeated wetting and drying
w	B	Development of color (higher chroma) or structure
x	B, C	Fragipan properties—firm (moist) with brittle failure when dry
y	B, C	Accumulation of gypsum
z	A, B, C	Accumulation of salts more soluble than gypsum

[a]   Master horizons to which the suffice is applicable.

followed by lowercase letters, and some A, E, and C master horizons are followed by lowercase letters.

3.5.1.1 Organic Layers in Well-Drained Soils Organic layers at the ground surface on well or seasonally drained soils are designated Oi, Oe, and Oa in *Soil Taxonomy* (Soil Survey Staff 1999), or L, F, and H in the Canadian and some other soil classification systems (Klinka et al. 1981). The L horizon is fresh or slightly decomposed plant detritus, or *litter*, the F horizon is *fermented* organic matter, and the H horizon is *humified* organic matter.

Most well-drained soils with forest or shrub plant communities have L or Oi horizons, some have F or Oe horizons, and fewer have H or Oa horizons. H horizons are common in cold forest soils, however. Poorly drained soils with Oi, Oe, and Oa horizons are very common in Alaska, Canada, and Florida. Well-drained soils with grassland or desert plant communities generally have no more than thin to discontinuous Oi (or L) horizons, if any O or L horizons.

A distinction is sometimes made between (1) soils with well-developed F, H, or F and H horizons and lacking thick A horizons, and (2) soils that have no more than thin L, F, and H horizons and much thicker A horizons. The former kind of organic matter distribution is called a *mor* humus type, and the latter is called a *mull* humus type. Intermediate between the two is a *moder* humus type. Soils with mor humus types are common in boreal conifer forests, and most soils in warmer climates have mull humus types.

3.5.1.2 Organic Layers in Wet Soils Canadian horizon nomenclature distinguishes between organic horizons that are saturated all or most of each year and those that are drained of excess water most of each year. The O horizons in the Canadian system correspond to those for wet soils in *Soil Taxonomy* (Soil Survey Staff 1999). Three kinds of O horizons are recognized (Oi, Oe, and Oa) by their visible fiber contents following rubbing between fingers. Oi horizons consist of fresh or only slightly decomposed organic matter with fiber > 40%. Oe horizons, with more decomposition, have 17 to 40% fiber. Oa horizons consist of greatly decomposed organic matter with fiber < 17%.

3.5.1.3 A Horizons   A horizons are inorganic layers that occur just below organic layers of inorganic, or mineral, soils. They are generally very dark greyish brown to black, because they contain more organic matter than lower horizons. Most soils, except those in recently exposed parent materials or arid climates, have A horizons. They are generally thicker in grassland soils than in forest soils because grasses have profuse root systems that contribute much organic matter to soils. Lowercase suffixes are seldom added to master A horizons.

3.5.1.4 E Horizons   E horizons are eluvial horizons bleached by the leaching of iron. Leaching of basic cations generally makes E horizons more acidic than horizons below them. Fine clay particles are leached from some E horizons also. E horizons generally form at the base of Oa horizons or below the A horizons. Well-developed E horizons are grey, or they assume the colors of insoluble minerals in the soils.

Leaching of iron and aluminum is accomplished by chelation of these elements in soils by mobile organic matter in Oa and strongly acidic A and E horizons. This process is most prevalent in cool, or boreal, forests. In poorly drained soils, iron can be leached following its reduction to the ferrous form, in which iron is more readily soluble than in its oxidized, or ferric, form.

3.5.1.5 B Horizons   B horizons are layers of slight (Bw) or intensive (Bo) alteration by weathering, shrink-swell, and other processes, or layers of illuviation. Illuvial materials that accumulate in B horizons can be of many different kinds. The most common illuvial horizons in humid temperate to tropical climates are Bt horizons of particulate clay accumulation. In cool, or boreal, forest, illuviation of aluminum and iron chelated with organic matter is common. These horizons are designated Bs if bright colors of iron compounds predominate, Bh horizons if black organic matter predominates, or Bhs horizons for those of intermediate character. In drier climates there are many different kinds of illuvial horizons. The most common are Bk horizons of calcium carbonate accumulation in semiarid areas to By horizons of gypsum accumulation in arid areas, and Bz horizons with soluble salt accumulation in poorly drained soils of dry climates. Bq horizons of silica accumulation are less common than Bk horizons, but they are common in semiarid areas with volcanic soil parent materials.

All B horizons take lowercase suffixes, and many have more than one suffix letter, for example, the previously mentioned Bhs horizon designation for a horizon with illuvial accumulations of both humus (h) and sesquioxides (s). When horizons become cemented, such as those with calcium carbonate or silica accumulation, a suffice m is added to call them Bkm or Bqm horizons.

3.5.1.6 C Horizons C horizons are layers of unaltered, or relatively unaltered, soil parent material. Obviously those C horizons that formed from the disintegration of bedrock must be somewhat altered by weathering. C horizons may have suffixes indicating accumulation of calcium carbonate or silica, for example, where that accumulation is considered to be more of a geological process than a pedological process.

3.5.1.7 R Layers R layers are hard bedrock. Weakly consolidated sedimentary rocks and rocks that are weathered soft (soft to a hammer), but are still very difficult to excavate with a spade, are designated Cr horizons. The interface between hard bedrock and soil is called a *lithic* contact, and that between soft bedrock and a soil is called a *paralithic* contact.

3.5.1.8 Soil Horizon Suffix Designations There are many different kinds of B horizons, and differences are indicated by the suffixes listed in Table 3.2. A and C horizons can have suffix designations also, but suffixes are not always added to A and C horizon designations.

3.5.2 Pedons and Soil Description

Individual plants and animals are described and compared in order to group them for classification and for transfer and application of knowledge gained from relatively few individuals to a large number of individuals in the same class. There are no discrete individuals in soils. The dimensions of individual soils are arbitrarily chosen to provide good representation of all horizons and be no larger than a volume that can be exposed and described in small excavations, or pits.

3.5.2.1 Pedon Definition The horizontal area of a pedon is arbitrarily chosen to be 1 m² (Figure 3.1), unless the soil has lateral variations

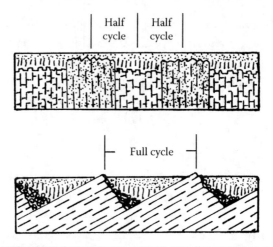

Figure 3.5   Cross sections of pedons in cyclic soils: Symmetrical cycles, with shallow soil with sandstone and deep soil with limestone parent material; asymmetrical cycles with alternate stony and nonstony soils.

that are repeated in cycles at intervals < 7 to 10 m across the landscape (Figure 3.5). In laterally cyclic soils, a pedon includes one-half of a cycle, and the maximum horizontal area is 10 m². If the lateral variations, or cycles, are asymmetrical, an entire cycle is necessary to describe all variation in a soil. Arnold (1964) argued that an examination of 0.8 (80%) of a cycle is commonly needed to describe 80 to 90% of the vertical variability. The vertical extent of a pedon is from the ground surface down to bedrock or unaltered regolith. According to *Soil Taxonomy*, the maximum depth of a pedon is 2 m, but soils are commonly described to greater depths. The evidence of roots and microorganisms can be found deeper than 2 m, even in mid-latitude areas (Richter and Markewitz 1995), and much deeper in tropical soils.

3.5.2.2 Pedon Description   Pedons are described in their natural landscape positions (Chapter 7). Laboratory data are supplemental to soil description and are seldom included in them. Soil color, mottles, texture (or field grade), structure, consistency, coarse fragment content, plant root distribution, and pH in indicator solution are generally included in pedon descriptions. Soil pH is estimated from a pinch of soil in an indicator solution, such as methyl red or bromthymol blue.

Interpedal voids that can be inferred from soil structure need not be described, but tubular and vesicular pores are described where present. It is commonly necessary to describe special features—those that do not occur in all soils—such as clay coatings (cutans), nodules, and effervescence in dilute HCl. According to tradition, soil depths are measured from the top of the mineral soil or, in organic soils, from the top of the organic soil. Officially, in the U.S. Department of Agriculture, soil depths are measured from the top of an Oi horizon, but the traditional method of measurement from the top of the mineral soil is more common in practice.

Pedon descriptions include landscape data that are pertinent to the site of a pedon. Most pertinent are landform, vegetative cover, and ground surface or slope characteristics. Although pedon excavations seldom penetrate bedrock, descriptions of bedrock fractures and weathering of bedrock beneath soil is important.

Problems

3.1. a. What is the exchangeable sodium percentage in a B horizon (subsoil) of Antioch clay that has 88 mmol+ of exchangeable Na^+, and exchangeable Ca^{2+}, Mg^{2+}, and K^+ of 156, 84, and 3 mmol+/kg, respectively, and lacks extractable acidity (no exchangeable H^+ or Al ions)?

Solution: ESP = 88/(88 + 156 + 84 + 3) = 26.6%

b. What is the sodium adsorption ratio of the soil?

Solution: SAR = $88/((156 + 84)/2)^{0.5}$ = $88/\sqrt{240}$ = 88/10.9 = 8.1.

Antioch clay has been classified as a Natrixeralf, indicating that it is a sodic soil with a natric horizon. It is the ESP, rather than the SAR, that identifies the natric horizon as such in the Antioch soil.

3.2. What is the mean particle density of a soil with 40% organic matter (Dp = 1.4 Mg/m^3) and 60% mineral (inorganic) matter?

Solution: 0.4 × 1.4 + 0.6 × 2.65 = 0.56 + 1.59 = 2.15 Mg/m^3

3.3. What is the porosity of a soil with a mean particle density of 2.6 Mg/m^3 and a bulk density of 1.5 Mg/m^3? What is the air-filled porosity of the soil when the water content is 18%?

Solution: P = 1 − Db/Dp
Total porosity = 1 − 1.5/2.6 = 1 − 0.58 = 0.42, or 42%
Air-filled porosity = 42 − 18 = 26%

3.4. What is the bulk density of a soil with a mean particle density of 2.2 Mg/m^3 and 42% total porosity?

Solution: P = 1 − Db/Dp
0.42 = 1 − Db/2.2
Db = 2.2(1 − 0.42) = 1.27 Mg/m^3

Questions

1. How are tubular pores formed in soils?
2. Why are roots more successful in growing through sandy soils with bulk densities between 1.4 and 1.75 Mg/m^3 than in growing through clayey soils of the same bulk densities?
3. Major, or master, soil horizons from A to D are assigned sequential letters, from tops to the bottoms of soil profiles. Why might E be an exception to this sequential designation?
4. The lower limits of deep pedons are somewhat arbitrary. What do you believe to be the lower limit for soils lacking bedrock (R or Cr horizon) within the 2 m depth? Should it be the lower limits of oxidation or reduction features? Or something else?

Supplemental Reading

Hillel, D. 1980. *Fundamentals of Soil Physics*. Academic Press, Orlando, FL.
IUSS Working Group. 2006. *World Reference Base for Soil Resources*. World Soil Resources Report 103. Food and Agricultural Organization of the UN, Rome.
Mitchell, J.K. 1976. *Fundamentals of Soil Behavior*. Wiley, New York.
Schoeneberger, P.J., D.A. Wysocki, E.C. Benham, and W.D. Broderson (compilers). 1998. *Field Book for Describing and Sampling Soils*. National Soil Survey Center, Natural Resources Conservation Service, Lincoln, NE.

Soil Survey Division Staff. 1993. *Soil Survey Manual.* U.S. Department of Agriculture Handbook 18. U.S. Government Printing Office, Washington, DC.

Soil Survey Staff. 1999. *Soil Taxonomy: A Basic System for Making and Interpreting Soil Surveys.* Agriculture Handbook 436. USDA, Washington, DC.

4

HEAT AND SOIL TEMPERATURE

Heat and temperature control the kinds and rates of physical and chemical reactions in soils. Most activities in soils are confined to a relatively narrow temperature range between the freezing and boiling points of water (0 and 100°C). Soil temperatures are generally much lower than the boiling point of water. They range up to about 80°C at the ground surface, or even higher on black soils exposed to direct solar radiation, with mean annual temperatures ranging up to about 30°C belowground.

4.1 Sources of Heat and Heat Transfer at the Surface of Earth

Earth is very hot at its centre, thousands of degrees Celsius. Heat is conducted upward by convection in the mantle. Loss of heat from the thinner ocean crust is much more rapid than from the continents. Much of the heat that flows through the crust is generated by radioactive decay, mainly from ^{238}U, ^{235}U, ^{232}Th, and ^{40}K (Li 2000). The temperature of continental crust increases an average of 2°C for each 100 m of descent below the surface of Earth (Condie 2005). Nevertheless, heat flow from the interior to the surface of Earth amounts to no more than 0.03% of the energy received from the sun. Some heat is produced by chemical and biological processes at the surface of Earth, but solar radiation is the major source of energy.

Solar radiation consists of electromagnetic waves. The energy in solar radiation is converted to heat when the radiation is absorbed by Earth; this is sensible heat, we can feel it. The sun and Earth emit radiation according to the principles of black bodies. Radiant energy from a black body is proportional to the fourth power of the absolute temperature of the body (Stefan–Boltzmann law). The sun, with a surface temperature of 5800 K, emits about 3×10^5 more radiant energy than Earth at 255K (Henderson-Sellers and Robinson 1986). Thus,

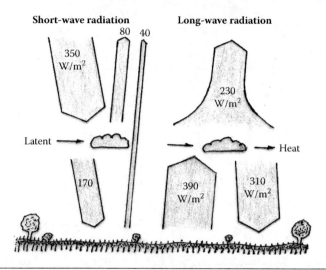

Figure 4.1   The radiation balance between short-wave radiation from the sun and long-wave radiation leaving Earth. This diagram does not show the quantities of latent heat transfers, nor conduction and convection transfers, which can have considerable lateral transfer components.

even though Earth is about 150×10^6 km from the sun, Earth receives enough energy from solar radiation to maintain a mean air temperature of about 14°C (287 K) at its surface. Solar radiation impinges on the outer atmosphere of Earth at a rate of 1350 to 1400 W/m². The mean incoming radiation at the top of the atmosphere over 24 h days for a year is about 400 W/m² at the equator and 200 W/m² at the poles (350 W/m² in Figure 4.1). Interception, reflection, and scattering of radiation in the atmosphere reduce the amounts of radiation that reach the ground surface by an average of about 100 W/m².

The frequency of radiation, which is the inverse of the wavelength, emitted from a black body is proportional to the absolute temperature of the body (Wien's law). Because the sun is about 23 times hotter than Earth, the frequency of radiation from the sun is much greater than that from Earth (Figure 4.2). Peak radiation from the sun is at a wavelength of 0.474 μm (or 474 nm). Our eyes are sensitive to radiation in the 400 to 700 nm range, which is called the visible spectrum, from violet to red. Much of the ultraviolet radiation, with wavelengths shorter than 400 nm, is absorbed by ozone in the stratosphere, and some of the infrared radiation, with wavelengths longer than 700 nm, is absorbed by water and carbon dioxide in the troposphere. Smaller amounts of infrared radiation are absorbed by nitrous oxides, oxygen,

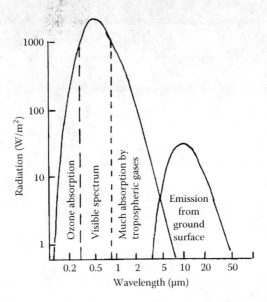

Figure 4.2 Wavelengths of radiation from the sun (5800 K) and radiation emitted from Earth (255 K).

methane, and clorofluorocarbon, which are less concentrated in Earth's atmosphere. Atmospheric scattering has a much greater affect on visible light than absorption, with the scattering being most effective with shorter-wave radiation; this is why the sky is blue (relatively short wavelength) until evening; in the evening, when the sun is low and the path of solar radiation through the atmosphere is much longer, only longer-wave radiation in the red end of the visible light range reaches the ground, making the sun and atmosphere appear red and orange to yellow.

Much of the solar radiation that passes through the atmosphere is reflected from clouds, vegetation, and soil or its cover of litter. The amount that is reflected from a surface is called the albedo of that surface; it ranges from < 10% for clear water to > 80% for clean snow and about 10 to 30% for soils (Table 4.1). Earth's albedo is greatest during winter, when there is snow in the northern hemisphere and many trees lose their leaves. The average albedo, or reflectivity from the surface of Earth, is 20 to 30% in tropical and temperate zones, increasing above 60° latitude to > 80% at the poles. Some energy absorbed by Earth is radiated back into the atmosphere. That outgoing radiation has much longer wavelengths than solar radiation and is more completely

Table 4.1   Short-Wave Reflectivity, or Albedo, of Natural Surfaces

Surface	Albedo (%)	Surface	Albedo (%)
Clouds		Soil	
Thickness < 15 m	5–60	Gray, dry	25–30
150–300 m thick	30–70	Gray, moist	10–12
300–600 m thick	60–80	Black, dry	14
Snow		Black, moist	8
Clean	80–95	White sand	35–40
Dirty	40–70	Grassland	10–30
Ice, clear	<10	Chaparral	15–20
Sea		Forest	
Smooth	7–8	Deciduous	10–20
Rough	12–14	Coniferous	5–15

Geiger et al. (2003) have many more examples.

absorbed in the atmosphere. For the surface of Earth to maintain its current mean temperature of about 14°C, incoming solar radiation must be balanced by outgoing radiation. Increases in atmospheric carbon dioxide and methane are expected to increase the amount of radiation that is absorbed in the atmosphere and radiated back to the ground, reducing the amount of loss and increasing the temperature of Earth's atmosphere and surface. This has been called the greenhouse effect.

Relatively small amounts of thermal energy are moved by wind and by the rise of warm air in the atmosphere. Besides the vertical transfers of radiant energy, wind transfers heated air laterally, and latent heat in the form of water vapor is transported from the subtropics toward both the tropics and higher latitudes, where the vapor in clouds condenses to water or ice. The transfer of latent heat is much greater than the transfer of heat in dry air by convection, because the volumetric heat capacity of air is small (Table 4.2). Latent heat transformations release a great amount of energy to the atmosphere—about 2497 J/g of water upon condensation and another 334 J/g when the water freezes, compared to 419 J released when 1 g of water cools 100°C from boiling to freezing temperatures. These heat transformations are reversed at the ground surface when ice and snow melt and water evaporates, cooling the atmosphere at the ground level. The greatest temperature effects of wind are to replace humid air at a surface, where water is being evaporated or transpired, with drier air, thus increasing the rates of evaporation or transpiration and the concomitant cooling.

Table 4.2 Thermal Properties of Rocks, Metals, and Substances in Soils

	Heat Capacity		Thermal Conductivity		
	Per Mass	Per Volume	0°	25°C or Unspecified	100°
Substance	kJ/kg K	MJ/m³ K		W/m K	
Granite	0.65	1.7	2.6		2.4
Gabbro	0.72	2.1	2.2		2.1
Basalt	0.85	2.2			
Dunite			5.2		3.9
Quartzite	0.70	1.9	6.2		5.2
Slate	0.71	2.0			
Silver	0.25	2.6		452	
Copper				384	
Aluminum				205	
Iron				72	
Water[a]	4.2	4.2		0.6	
Ice[b]	2.1	1.9		2.23	
Air	1.0	0.0013		0.02	
Dry sand	0.8	1.2		0.16	
Wet soil	1.5	2.7		1.7	
Galveston Clay					
Dry				0.24	
Wet				1.5	

Source: Rosenberg, N.J. et al., *Microclimate, the Biological Environment*, Wiley, New York, 1983; Goranson, R., *Geological Society of America*, Special Paper, 36, 223–242, 1942; Birch, F., *Geological Society of America*, Special Paper, 36, 243–266, 1942; Hillel, D., *Fundamentals of Soil Physics*, Academic Press, Orlando, FL, 1980.

[a] Latent heat of vaporization ranges from 2497 kJ/kg at 0°C to 2261 kJ/kg at 1000°C.

[b] Latent heat of melting is 334 kJ/kg.

4.1.1 Variations in Energy Flow to and from Earth

Earth climate has fluctuated widely from the ice-free Eocene more than 36 million years ago to Quaternary ice ages over the last 2 to 2.4 million years. There are cyclical changes attributed to differences in the amounts or timing of solar radiation reaching Earth and noncyclical changes related to the composition of Earth's atmosphere.

Wide temperature fluctuations during the Quaternary have been attributed to the distance of Earth from the sun and the orientation

of Earth with respect to the sun. The eccentricity of Earth's elliptical orbit around the sun and rotation of the tilting Earth axis, which
is currently tilted at 23.4°, cause approximately 100, 41, and 21 ka
fluctuations in climate (Briggs et al. 1997). For the last 12 or 15 ka
Earth has been in a relatively warm stage called the Holocene. Within
the Holocene variations in solar activity have caused slight fluctuations in solar radiation on approximately 11- and 22-year cycles. The
impacts of fluctuations in solar activity, or sunspots, on Earth climate
are controversial.

The noncyclic, or less predictable, fluctuations in Earth climate
are either cooling by reflection of incoming short-wave radiation by
aerosols or clouds, or warming by increased absorption of mainly
long-wave radiation by atmospheric gases. Losses of polar ice will
affect the albedo of Earth greatly.

Violent volcanic eruptions that add particulate aerosols and SO_2
to the atmosphere can cause minor cooling for a few days to major
cooling for a year or more, as in the case of the Krakatoa eruption
in 1883, when the sky was darkened for a long time. The particulate
aerosols reflect and scatter incoming short-wave radiation, reducing
the amount that reaches the ground, and the SO_2 becomes hydrated
to form H_2SO_4 aerosol that promotes cloud formation. It has been
proposed that successive volcanic eruptions over many years during
production of the massive Siberian Trap basalts caused extensive cooling that resulted in the great extinction of life and loss of entire species
at the end of the Paleozoic era about one-quarter of a billion years ago.

The main atmospheric gases that absorb the most outgoing radiation and radiate it back to the surface of Earth are H_2O, CO_2, CH_4,
N_2O, and chlorofluorocarbons. Carbon dioxide has had the greatest
effect, because it is the most abundant of these greenhouse gases, other
than H_2O, and it has fluctuated widely. Atmospheric CO_2 concentration was much greater than currently during the warm Eocene, lower
during Pleistocene glacial stages, and has been rising steadily from
0.0318% when first measured at the Mauna Loa Observatory in 1958,
and it is nearing 0.04% of the gas content in the atmosphere (Chapter
11). Because of large climatic differences from year to year, it is difficult to quantify trends, but a warming trend is evident. The warming
trend caused by increased CO_2 and methane from the burning of coal,
hydrocarbons, and vegetation would be much greater, except that the

burning releases aerosols and SO_2 that counter warming by their cooling effects. Opening of the Arctic Ocean by early melting of the ice to allow the passing of ships during summers should be enough evidence to convince sceptics of global warming who do not want to rely on the climatic data. The surest consequence of global warming is the rise of sea level as polar ice melts and as water in the oceans expands as it becomes warmer. Global precipitation will increase as more water evaporates from the warmer oceans, but it is uncertain where the extra precipitation will fall; some areas will become wetter and others will become drier. Strom (2007) provides a good assessment of global warming and its possible consequences.

4.1.2 Latitudinal and Slope Effects

The axis of Earth is tilted 23.4° from the plane of its orbit around the sun. Therefore, midday radiation from the sun falls vertically upon 23.4°N at the summer solstice, about June 21 to 22, vertically upon 23.4°S at the winter solstice, about December 21 to 22, and upon the equator during spring and autumn. Two rainy seasons are common in subtropical areas, related to the fact that the midday sun is directly overhead at two different times during the year.

The maximum proportion of solar radiation that is incident upon a sloping surface is proportional to the cosine of the angle between a line to the sun and a line that is perpendicular to the slope. That amount is 100% of the solar radiation when the path of radiation from the sun is perpendicular to the slope, or down to 0% when the sun cannot be seen over the slope horizon. At midday the maximum direct radiation falling on a horizontal (tangent) surface at the north poles, which is at the summer solstice, is 40% of the amount at 23.4° latitude when the sun is directly overhead there. The direct radiation at the poles is actually less than 40%, because of greater diffusion and absorption of radiation on a longer path through the atmosphere to the poles. On the longest polar days of 24 hours, the total daily radiation in the atmosphere above the pole is slightly greater than that at 23.4° latitude if atmospheric diffusion and absorption are ignored (Briggs et al. 1997).

Slopes facing toward the equator receive more direct solar radiation than level surfaces, and steep slopes at 45 to 60° latitude receive more than twice as much direct solar radiation as pole-facing slopes at these latitudes. Much of the solar radiation that reaches the ground is scattered or diffuse radiation; however, that radiation falls equally on pole-facing slopes and on those facing toward the equator. At higher latitudes diffuse radiation can be 30 to 40% of the total incoming solar radiation.

4.1.3 Ground Cover and Soil Color

Solar radiation that passes through the atmosphere is reflected or absorbed by vegetation, plant detritus, or bare soil. Radiation that reaches the soil and is not reflected back to the atmosphere is absorbed within a few millimeters of the surface. Soil color and its reflectivity are important when there is incomplete ground cover. Reflectivity from bare pumice leeward (eastward) from Mt. Mazama, now occupied by Crater Lake, is so great that the energy loss during cold dry weather causes severe frost that limits the regeneration of yellow pine trees. This hazard is mitigated on darker-colored soils that absorb more solar radiation and radiate some of the energy back to the atmosphere at night.

4.1.4 Heat Exchange in Soils

Heat exchange and temperature differences within soils are dependent on heat storage and conductivity and on phase changes of water. Water holds more heat than other substances, but heat flow is more rapid through rocks and wet soil (Table 4.2). Heat flow (q) is proportional to the temperature gradient (∇T) and the thermal conductivity (k) across the gradient:

$$q = -k\,(\nabla T), \text{ Fourier's law}$$

The high heat capacity of water dampens temperature changes in soils. The melting of ice and the vaporization of water further dampen temperature changes, because both phase changes take large amounts of heat (Table 4.2, footnotes). Conversely, heat is released in the condensation of vapor and in the freezing of water. Horticulturalists utilize this phenomenon by spraying water in citrus orchards on cold nights, expecting that heat released by freezing water will prevent or mitigate damage to foliage and fruit.

Air has a greater effect on heat transfer than its low conductivity would indicate, because air transports heat by convection, as well as by conduction. Nevertheless, the role of air in heat transfer within soils is relatively small.

4.2 Soil Temperatures

Soil temperatures fluctuate through daily (diurnal) and annual cycles (Figures 4.3 and 4.4) and differ in response to changes in vegetative and litter cover (Figure 4.5). Daily soil temperature fluctuations are greatest at the ground surface and diminish exponentially with depth (Smith et al. 1964). The fluctuations are greater in bare soil than in soil covered by vegetation and litter. Plants shade the ground from the sun, and a layer of litter traps air, a poor conductor of heat, near the ground surface. Surface soil temperatures in Minnesota during July range from 18 to 34°C under bare ground and only from 19 to 26°C under sod, although the monthly means are less extreme (Figure 4.3).

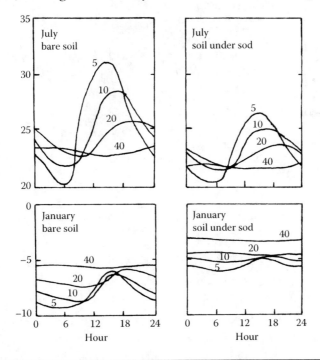

Figure 4.3 Daily cycles of soil temperatures in a Minnesota soil at 5-, 10-, 20-, and 40-cm depths under bare ground and under sod during January and July. (Data from Baker, D.G., *Minnesota Farm and Home Science*, 22(4), 11–13, 1965.)

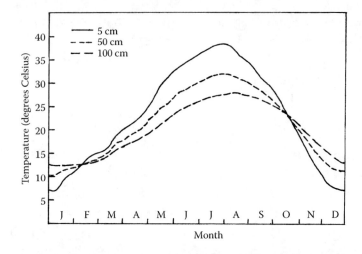

Figure 4.4 Soil temperatures at Fresno, California, averaged over 10 years (1969–1978). (Data from Fox, J.A., and J.L. Hatfield, *Soil Temperatures in California*, Bulletin 1908, University of California, Division of Agricultural Sciences, 1983.)

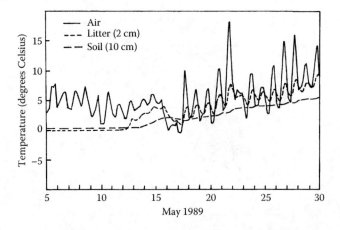

Figure 4.5 Soil temperatures at Kowee Creek in southeast Alaska from May 5–30, 1989: temperatures aboveground, beneath 2 cm of plant detritus (litter), and at 10 cm depth in soil.

Daily soil temperature fluctuations are very small or negligible below about 0.2 m in dry soils and below 0.5 m in wet soils.

Well-drained soils that are sandy warm up more rapidly than clayey soils during the spring, because they hold less water. Assuming that the soils were saturated during the winter, clayey soils conduct heat more rapidly than drained sandy soils, but it takes much more heat to warm them because of the large heat capacity of water. Annual cycles of heating and cooling affect soil temperatures to depths of 5 m or greater. Mean annual temperatures, however, are nearly the same at all depths, down several meters in most soils and down hundreds of meters in deeply frozen soils. At greater depths belowground, temperatures rise about 1 or 2°C for each additional 100 m of depth, ranging up to 4 or 5°C per 100 m, where heat flow from the mantle through the crust of Earth is greatest.

Soil temperature differences related to the slope aspect are greater at higher latitudes, but are evident as low as 14°N latitude (Alexander 1976). At Tegucigalpa, aspect-related differences are barely noticeable near the summer solstice when the sun is nearer the zenith, and greatest near the winter solstice when the sun is lower in the sky. The warmest slope direction is south during April, which is late in the dry season, and southeast during the wet season, when cloudiness and rainfall are prevalent in afternoons. More commonly, southwest-facing slopes are warmer than southeast-facing slopes in the northern hemisphere.

Busse et al. (2010) recorded soil temperatures at 2.5-, 5-, 10-, and 15-cm depths in forest soils that were burned over with heavy fuel loads of masticated understory trees and shrubs. The temperatures were about the same in both sandy loam and clay loam soils. Maximum temperatures at the 2.5-cm depth were > 200°C on dry disturbed soils and < 200°C on dry undisturbed soils and were > 100°C on moist disturbed soils and < 100°C on moist undisturbed soils. Temperatures > 60°C had been found to kill the plant roots. Lethal temperatures occurred down to 5 cm in moist soils and down to 10 cm in dry soils.

4.3 Soil Temperature Regimes

Soil temperature is a property that is utilized in the classification of soils (Soil Survey Staff 1999). There are nine soil temperature regimes (STRs): pergelic, cryic, frigid, isofrigid, mesic, isomesic, thermic, isothermic, and hyperthermic. Some STRs are defined entirely and others partly, by mean annual soil temperatures measured at 50-cm depth, or at the top of bedrock if that is within 50 cm of the ground surface. The major STRs and the limits between them are pergelic < 0°C > frigid < 8°C > mesic < 15°C > thermic < 22°C > hyperthermic. Soils in isoclasses have mean summer and winter temperatures that differ by no more than 5°C; for example, a soil with a mean summer temperature of 16°C and a mean winter temperature of 12°C (mean, 14°C) is in an isomesic soil temperature regime. Cryic soils have mean annual temperatures in the frigid range (0 to 8°C) but remain cooler than frigid soils through summers.

The soil temperature regimes have some practical implications. For example, many crops, such as corn and winter wheat, are unsuited for frigid and colder regimes, and banana plantations are confined to hyperthermic regimes. Soil temperature regimes are mapped by extrapolation from relatively few sites with complete data, because soil temperatures may differ so much from year to year that several years of measurement are necessary to obtain reliable averages. The extrapolation of soil temperature regimes is generally inferred from topography, altitude, and local plant species distributions.

4.4 Soil Temperature Effects

Heat and soil temperatures affect soils and soil processes both directly and indirectly. The main indirect effects are freezing and evaporation of water.

Water expands when it freezes, pushing soils apart and damaging plant roots. Water freezing in cracks in rocks can cause them to disintegrate. The greatest effects of freezing in soils are in polar areas where stones are sorted from soils and distinctive ground patterns develop. Most soils with pergelic soil temperature regimes are frozen, but some are dry, particularly in Antarctica. The ice in the surfaces of frozen soils melts during summers, causing them to be wet, because subsurface ice restricts drainage of water from the soils. Thus, wetlands are extensive in polar and subpolar areas. The thawed surface in soils with permanently frozen subsoils, or *permafrost*, is called the *active layer*. In frigid soils, the detrimental effects of freezing are greatest in soils with groundwater near enough to the ground surface to replenish water at the freezing front as water is concentrated there by freezing. Freezing in soils with limited water supply causes desiccation, which can result in contraction or collapse of soil from which water is drawn to the freezing front.

Both low soil temperatures, near freezing, and high soil temperature ($T > 30°C$) severely restrict plant root growth and animal activity.

Curtailment of microorganism activity at lower temperatures retards the decay of organic matter and allows it to accumulate in soils. In general, soil organic matter is very beneficial for plants, but with very rapid accumulation of organic matter, most of the plant nutrients are sequestered in organic matter and insufficient nutrients are released by decay to furnish an adequate supply of nutrients for optimum plant growth.

Slope aspect differences are much easier to observe in vegetative cover than in soils, but the effects on soils can be substantial also. One of the many studies of soil differences related to slope aspect was in southeastern Ohio, where the soils in oak and hickory forest on southwest-facing slopes have more strongly developed subsoils (B horizons) than soils in mixed hardwood forest on northeast-facing slopes (Finney et al. 1962). Slopes are commonly steeper, and the soils deeper on north-facing slopes than on south-facing slopes in the northern hemisphere, although this is not a readily obvious trend (Alexander 1995), and the vegetation is commonly more xerophytic on south-facing slopes. Krause et al. (1959) reported that at a location in central Alaska, at about 65°N latitude, soils on south-facing slopes were well drained and those on north-facing slopes were poorly drained soils with permafrost.

Note on Units of Measure

The most familiar unit of heat, or thermal energy, is the calorie (cal). It is defined as the amount of energy, or heat, required to raise the temperature of 1 g of water by 1°C (or K). The SI unit of energy is the Joule, which is 4.19 calories, rounded to 4.2 in Table 4.2.

For solar radiation, the langley (cal/cm^2) has been replaced by Joules per square meter. One J/m^2 equals 4.19×10^4 langley. The unit for rate of energy flow, or flux, is the watt (W), which is a Joule per second (J/s).

Questions

1. Why does the sun appear redder at sunset than in the middle of the day?
2. Will loss of polar ice increase or decrease the albedo of Earth? Why?

3. How was it determined in Section 4.1 that direct solar radiation at the north pole at midday on the summer solstice is 40% of that at 23.4° latitude, where the sun is directly overhead?

Solution: The angle from the pole to the sun when it is overhead at 23.4° latitude is 90 − 23.4, or 66.6°. The cos of 0° is 1, and the cos 66.6° equals the sin 23.4° (or 23°24'), which is 0.39715, or 40% of 1 (or 100%) at 23.4° latitude.

4. Is diffuse solar radiation greater on slopes facing toward the north pole or on slopes facing southward toward the equator?
5. Are daily soil subsurface (below 50 mm) temperature fluctuations greater in dry or wet soils during summer?
6. In the experimental burning of Busse et al. (2010), why were the dry soils heated to higher temperatures and to greater depths than the moist soils?

Supplemental Reading

Condie, K.C. 2005. *Earth as an Evolving Planetary System*. Elsevier, Amsterdam.
Geiger, R., R.H. Aron, and P. Todhunter. 2003. *The Climate Near the Ground*. Rowan and Littlefield, Lanham, MD.
Hillel, D. 1980. *Fundamentals of Soil Physics*. Academic Press, Orlando, FL.
Rosenberg, N.J., B.L. Blad, and S.B. Verma. 1983. *Microclimate, the Biological Environment*. Wiley, New York.
Smith, G.D., F. Newhall, and L.H. Robinson. 1964. *Soil Temperature Regimes—Their Characteristics and Predictability*. SCS-TP-144. USDA, Soil Conservation Service, Washington, DC.
Soil Survey Staff. 1975. *Soil Taxonomy*. U.S. Government Printing Office, Washington, DC.
Strom, R. 2007. *Hot House: Global Climatic Change and the Human Condition*. Copernicus Books, Springer, New York.

5

SOIL WATER, AIR, AND CLIMATE

Water is essential for the growth and survival of living organisms. Few vascular plants can survive solely on water taken directly from the atmosphere. Most of the water that these plants use comes from the soil, through their roots. Soil water supply is a limiting factor in most terrestrial ecosystems. Therefore, a consideration of soil properties or characteristics that affect the supply of water to plant roots is essential in assessing ecosystem productivity.

Water in precipitation that falls on the ground enters soils or flows overland to streams or ponds. The larger portion of it, that which enters soils, flows through soils to a groundwater reservoir or is stored until it is extracted by plants or returned to the atmosphere by evaporation (Figure 5.1).

The supply of water to plants depends on the amount of precipitation, the proportion of that precipitation that enters the soil, the amount of water that can be stored in a soil, transport of water to plant roots, and the energy required to take water from the soil into roots. Major soil properties and characteristics that affect the supply of water to roots are the rate of infiltration of water into the ground, movement of water within and between soils, storage capacity for water within the soil, movement of water to plant roots, and solute content of water in the root zone.

Some soils contain too much water for many plants. When soils are saturated there are no air-filled pores and the exchange of gases is very slow, causing oxygen to be depleted in root zones. Only specialized plants that are adapted to saturated soil conditions are able to take up all of the nutrients required for growth and survival in soils with scarce oxygen.

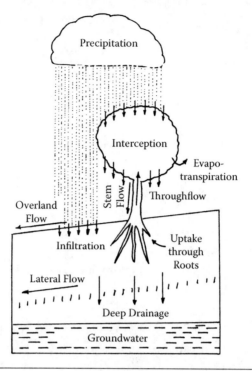

Figure 5.1   The flow of water in a soil landscape.

5.1 Infiltration of Water into Soils

Most water enters soils from rain or melting snow, which together are called precipitation. Some precipitation is intercepted and retained by plants, but most of it falls to the ground, drips through plants, or flows down stems to the ground (Figure 5.1). Most water that reaches the ground *infiltrates* into soils. Water that infiltrates into soils is either retained in them or flows through to other soils in lower topographic positions or to a groundwater table.

Infiltration is promoted by plant detritus, or litter, that absorbs raindrop impacts and retards overland flow of water. Where the ground is not protected by plant litter or rock fragments, raindrop impact on wet soils destroys aggregation, and splash from saturated ground carries soil particles downslope. Loss of aggregation in wet soils is called *puddling*. In puddled soils, the larger pores are filled with finer particles, plugging pores that might otherwise carry water through soils. This reduces infiltration and increases the likelihood that some water will flow overland, carrying soil particles with it. Soil loss by overland flow, or erosion, is negligible in most undisturbed plant communities absent exceptional rainfall or snowmelt events.

Surface soils in forests and shrublands that become hot and very dry during summers can become hydrophobic and repel water (DeBano 2000). Hydrophobic soils can retard infiltration until the surface soils are soaked long enough to allow water to pass through unimpeded. Diminished infiltration can cause overland flow of water and potentially soil erosion. Although hydrophobity can develop quickly, especially in wildland fires, it is ameliorated naturally.

5.2 Energy Status and Disposition of Water in Soils

The movement of water in soils and its availability to plants and other living organisms is dependent on the pressures and stresses (negative pressure) of the water. The water status of a soil is defined by the amount of water held and by the energy with which it is held relative to a free water surface at atmospheric pressure. The energy is referred to as a *potential*, because the *kinetic* energy of water in soils is negligible. Sources of the potential energy are gravity (gravitational potential), adsorption of water on soil surfaces and capillary forces (matric potential), and solutes in water (osmotic potential). Soil water potentials are positive only in saturated soils in which the pressure is greater than atmospheric pressure. In saturated soils, water flows downward under the influence of gravity, unless it encounters an impermeable barrier or a groundwater table. In unsaturated soils, the water commonly flows slowly toward roots, where water is being depleted from soils by plant uptake of water.

Because the water pressures and potentials in soils that are not saturated with water are generally negative, it has been convenient to utilize *suction* (negative pressure) in discussing soil water, but that practice has been discontinued by soil physicists. Pressure is force per area; therefore, the SI units are Newtons per square meter (N/m^2), or pascals (Pa). A pressure of 1 bar, which is approximately 1 atm, is equivalent to 10^5 Pa, a very large number; therefore, it is more convenient to express soil water potentials in kilopascals (kPa) or megapascals (MPa). Energy is force times distance, or pressure times volume. For a unit volume of water, or a unit mass, because water is practically incompressible, energy is equivalent to pressure; for example, 1 J/kg = 1 kPa.

5.3 Water Storage and Air in Soils

5.3.1 Field Capacity

Gravity causes water to enter soils and flow through them. If there are no soil drainage restrictions, only water that is held more strongly than it is attracted by gravity will be retained in soils. Capillary forces, which are due to a combination of adhesion of water to soil surfaces and surface tension caused by intermolecular forces in water, counterbalance the force of gravity and retain water in pores smaller than about 30 μm diameter. Thus, water in pores smaller than about 30 μm will be retained, and most water in larger pores will be lost from freely drained soils. Some water will be retained in larger pores that have exits through capillary pores that restrict drainage from the larger pores.

The water content of a soil that has been saturated and then allowed to drain freely for a day or two is called *field capacity*. It is a somewhat arbitrary soil characteristic that is approximated by applying a suction of 0.1 bar (–0.01 MPa or –10 kPa pressure) to a core or clod sample of soil and allowing the soil to drain until it is equilibrated with a water stress of –10 kPa. A suction of 0.3 bar (–30 kPa pressure) has been utilized traditionally in the United States to approximate the field capacity of agricultural soils that are not sandy, but a suction of 0.06 bar (–6 kPa pressure) is assumed in many countries.

5.3.2 Wilting Point

Plants extract water from soils at stresses up to –1500 kPa (–1.5 MPa) or greater. A water potential of –1.5 MPa is considered to be the wilting point. Most plants derive no benefits from water at greater stresses other than survival, rather than wilting and dying. Water held at a potential of –1.5 MPa is in very fine capillary pores (diameter < 0.5 μm) or adsorbed on mineral and organic matter (OM) surfaces. Because most of the surface area and very fine capillary pores in soils are attributable to clay size particles, amorphous materials, and organic matter, the amount of water held at the (permanent) wilting point is an index of the amounts of these components in soils. Clay holds about 0.4 g of water (only slightly more for smectite and slightly less for kaolinite) per gram, and humus about 1.5 g of water per gram at –1.5 MPa pressure. Different kinds of amorphous materials hold different amounts of water, but generally more than 0.6 g of water per gram of amorphous material at the wilting point. In many soils, which have little amorphous material and no more than 1 or 2% humus, the clay content can be approximated by multiplying the –1.5 MPa water content by 2.5 (Soil Survey Staff 1999); for example, a subsoil with 6% water held at –1.5 MPa might by expected to have about 15% clay. If that soil is analyzed for clay content and is found to have only 8% clay, rather than 15% clay, amorphous materials are expected to be important components in the soil.

5.3.3 Saturation

When all pores in a soil are occupied by water, it is *saturated* with water. The amount of water at *saturation*, or the total storage capacity of a soil, is equal to its total porosity. Soils are completely saturated only after an intense rain, under melting snow, below the groundwater table, or where a drainage restriction such as a claypan or a hardpan ponds water above the groundwater table.

5.3.4 Aeration Porosity

When saturated soil drains, the pores that are vacated fill with air. The pore space filled with air at field capacity, or the difference between saturation and field capacity, is sometimes referred to as the *aeration porosity* or *air capacity* of a soil.

5.3.5 Available Water Capacity

The plant available water capacity (AWC) of a soil is the amount of water held between field capacity (–0.01 MPa) and the permanent wilting point (–1.5 MPa). It is hardly important that some plants extract water at stresses considerably greater than –3 MPa, because soils generally hold little more water at –1.5 MPa, which is the official

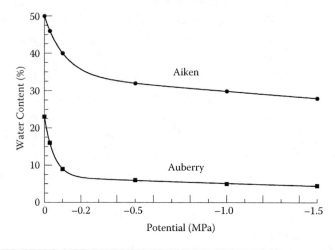

Figure 5.2   Soil water retention curves (gravimetric water) for two California soils. The wilting point and field capacity of Aiken are high for a clay loam, because the soil contains appreciable amorphous material.

wilting point, than at −3 MPa. The stress assumed for field capacity is much more critical than that assumed for wilting point; for example, the difference between −0.01 and −0.03 MPa in the Auberry sandy loam is 7% water, or about 3/8 (38%) of the −0.01 to −1.5 MPa water retention, or AWC (Figure 5.2).

The capacities of soils to hold plant available water range widely. Assuming that field capacity is at −0.01 MPa stress, then the volumetric AWC of soils ranges from < 4% in coarse sandy soils to > 21% in ashy (volcanic) soils (Table 5.1). Soil organic matter in inorganic soils increases the gravimetric AWC substantially, but it has little effect on the volumetric AWC of inorganic soils, except sandy soils, because higher gravimetric water holding capacity due to organic matter is counterbalanced by greater volume per mass (or lower bulk density) of soil containing organic matter (Problems 5.1 and 5.2).

AWC is generally reported on a volumetric basis so that the capacity of a soil to hold plant available water in a given thickness or depth can be readily computed from a soil description containing soil textures (Problem 5.3). The thickness of each soil layer is multiplied by its AWC (Table 5.1), divided by 100, and the results for all layers are added together. The AWC of a layer with stones is reduced by the volume of the stones, unless the stones are porous and hold known amounts of available water. In deep soils, the total amount of plant

Table 5.1 Volumetric Available Water Capacities of Mineral Soils by Field Grade, or Soil Textural Class, Based on Water Retention between 0.1 and 15 Bars of Suction

Field Grade	AWC[a] (%)	Field Grade	AWC[a] (%)	Field Grade	AWC[a] (%)
Ashy[b]	>21	Sand		Sandy loam	15
Clay		Coarse sand		Coarse	12
Coarse	16	OM < 1%	4	Fine	18
Fine (clay > 60%)	12	OM > 1%	6	Silt loam	
Clay loam	19	Fine sand		Coarse	24
Loam	21	OM < 1%	8	Fine (clay > 18%)	22
Loamy sand		OM > 1%	10	Silty clay	15
OM < 1%	10	Sandy clay	14	Silty clay loam	18
OM > 1%	12	Sandy clay loam	15		

[a] The AWC in percent (volume %) divided by 100 gives the AWC in cm/cm or in./in., so that the available water retained in a column of soil is readily computed from the field grade or textures in a soil description.

[b] The AWCs of soils dominated by volcanic ash and cinders are highly variable and unpredictable. AWCs are generally >21% and can be more than double this lower limit.

available water is dependent on root depth and density, which are different for different soils and different kinds of plants. Root densities generally decrease at greater depth, so plant available water declines with soil depth even within the root zone. Because plant root densities are very difficult to estimate, AWCs are generally computed for the upper 1 or 1.5 m of soil, without regard to root density.

5.3.6 Solute, or Osmotic, Potential

Water containing solutes has a negative potential that decreases (becomes more negative) as the solute content increases. It is negative because water will flow across a membrane that is permeable to water and impermeable to solutes, from a solution that is less concentrated (lower stress or osmotic potential) to one that is more concentrated (greater, more negative osmotic potential) in solutes. A plant expends more energy in taking water from a salty soil than from a nonsalty soil, and the vapor pressure is lower and evaporation slower from more salty soils. The solutions in most soils of nonarid areas are dilute at field capacity, increasing in concentration as soil water is lost. Salt concentrations in extracts from saturated saline soils are on the order of 0.1 to 0.2 molar. When the water content in a salty soil is at or less than field capacity, plants require a potential equal to the sum of the matric and osmotic potentials to acquire water from the soil. Saline soils are generally in arid areas, closed basins in semiarid areas, and adjacent to sea coasts.

5.3.7 Some Management Implications

The optimum soil water condition for the growth of most plants is in the range from field capacity to −0.1 or −0.2 MPa of pressure (Table 5.2). This condition allows movement of air through macropores, and plants acquire water with minimal expenditure of energy. Aeration is generally best in sandy soils and in soils with granular structure, but sandy soils hold less water available for plant growth and survival (Figure 5.3).

The kinds of soils that provide optimum water retention for plant growth are different in different climates. In a climate with frequent rain, a sandy soil is optimum, because drainage is rapid and aeration porosity is high. In most climates, however, the precipitation is so infrequent that a sandy soil would dry below an optimum water content between at least some rain events; therefore, a soil with some silt and clay is generally optimal, for example, a loam.

If most of the precipitation is during winters when there is little evapotranspiration, and the precipitation is enough to saturate loam

Table 5.2　Soil Water Disposition and Plant Condition or Growth

Moisture Condition	Soil Water Potential		Soil Water Movement	Plant Condition
	Bars	MPa		
Saturation	< −0.1	< −0.01	Flow in macropores due to gravity	Anaerobic environment restricted root growth
Field capacity	−0.1 to −2	−0.01 to −0.2	Capillary flow due to surface tension caused by intermolecular attraction in water	Unlimited growth
	−2 to −6	−0.2 to −0.6		Reduced transpiration and growth
	−6 to −15	−0.6 to −1.5		Survival, but no growth
Wilting point	−15	−1.5	Vapor diffusion due to kinetic energy of molecules and vapor convection due to pressure gradients	Dessication and death
Hygroscopic moisture	−15 to −10^4	−1.5 to −10^3		
Oven dry	−10^4	−10^3	No water	No plants

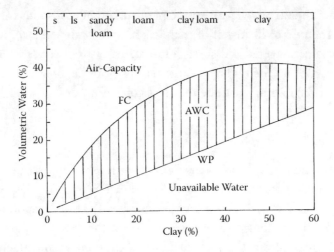

Figure 5.3 The wilting point (WP), field capacity (FC), and plant available water capacities (AWCs) of a sequence of soils with textures from sand (s) to loam to clay. The soils are assumed to have little organic matter and bulk densities of 1.2 Mg/m³; the solid phase is 45% of the soil volume and the total porosity is 55%.

or clay loam soils, then these are the ideal textures, because they hold the most plant available water (Figure 5.3). If most of the precipitation is in small increments during the growing season, then a sandy soil may be ideal. Compare loamy sand and clay loam soils 30 cm deep, for example: A rainfall of 3.6 cm on dry loamy sand would raise the moisture level of the entire soil to field capacity, and 3.0 cm of water would be available to plants. The same amount of rainfall on a dry clay loam would raise the moisture of the upper 9 cm to field capacity, or a greater depth to less than field capacity, and the maximum amount of water available to plants would be no more than 1.8 cm.

5.4 Movement of Water and Air in Soils

Water in soils moves from areas of higher potential to areas of lower potential. In saturated soils it moves through macropores in response to pressure differences. In unsaturated soils it moves in capillary pores and as a vapor phase in macropores. Movement is most rapid in saturated soils and declines as water in the larger pores is replaced by air.

Soil air also moves through macropores in response to pressure gradients, but the movement of gases in soil air is largely by diffusion from areas of higher to lower concentration. Water moves in soil air in the vapor phase by the same principles that other gases move through soils. The rate of diffusion of each gas in soil air depends on its concentration, or partial pressure, gradient, and is independent of other gases in the air. The vapor pressure in moist soils depends on the water potential, including that due to solute concentration.

5.4.1 Soil Permeability

The permeability of a soil is its facility to transmit liquids and gases. It depends on the amount and nature of soil porosity. The pore size distribution and their continuity and tortuosity are all important features. Friction along pore walls and turbulence caused by pore wall irregularities retard the flow of fluids. Large pores are much more effective conduits for fluids than smaller ones, because the ratios of cross-sectional area to perimeter are much greater. Large pores are not effective, however, if they are discontinuous. A soil with many vesicular pores formed by bubbles of gas can have a high porosity and low permeability; among rocks, rhyolitic tuff is a similar example. In soils, vesicular pores are common only near the ground surface.

5.4.2 Saturated Hydraulic Conductivity

The hydraulic conductivity of a soil is its facility to transmit water. Saturated hydraulic conductivity (K_{sat}) is the rate that water flows through a saturated soil or bedrock from a position of positive water potential. It ranges from very rapid (very high) in basalt, weathered limestone, some sandstones and conglomerates, and gravelly sands to very slow (very low) in most rocks and clayey sediments and clayey soils (Table 5.3). Crude estimates of K_{sat} can be made from soil texture and structure. The effect of texture is obvious—coarser-textured soils have higher conductivities. Dense massive soils and soils with platy structure have low conductivities, and soils with granular or strong blocky structures have higher conductivities. Soils with prismatic structure may have high permeabilities when prisms are dry and water can drain through vertical cracks between prisms, becoming low when prisms expand upon wetting to close pore spaces between prisms. High Na contents in soils with enough bicarbonate to raise the pH above 8.4 can cause dispersion of clays and greatly reduce the hydraulic conductivity.

Table 5.3 Saturated Hydraulic Conductivity of Soils

Conductivity Class[a]	Conductivity Rate		Kinds of Soil	
			Massive or Platy	Open Structure[b]
	μm/s	cm/h	Texture[c]	
Very high	>100	>36	Very gravelly sand	Coarse sand
High	10–100	3.6–36	Gravelly sand	Sandy loam, fine sand
Moderately high	1–10	0.36–3.6	Sandy loam	Loam, clay loam, silt loam
Moderately low	0.1–1	0.036–0.36	Sandy clay loam, loam, silt loam	Clay, silty clay loam
Low	0.01–0.1	0.0036–0.036	Clay loam, silty clay loam	Fine clay
Very low	<0.01	<0.0036	Clay	

[a] Class names and limits from *Soil Survey Manual* (Soil Survey Division Staff 1993).
[b] Granular or strong, fine blocky structure, low bulk densities, or soils with many biopores (pores made by plant roots and small animals).
[c] Exhibit 618-9 in *National Soil Survey Handbook* (Soil Survey Staff 1993) contains more detail.

Quantitatively, the rate of flow (q) in saturated soils follows Darcy's law:

$$q = K_{sat} (\Delta H/L), \text{Darcy's law}$$

where ΔH is the head, or elevation difference, and L is the length of the path from the higher to the lower elevation. The $\Delta H/L$ term is called the *hydraulic gradient*. The direction of flow is generally assumed to be downward, but it can be upward with artesian conditions (next section). The pore sizes and their distributions in soils are so differential that much of the water that drains through soils flows through relatively few of the larger pores. The rate that water drains from a saturated soil depends on the rate of flow through the layer with the lowest hydraulic conductivity (Problem 5.4).

5.4.3 Direction of Flow

Water in saturated soils flows downward in response to gravity. When water flowing downward encounters a much less permeable layer of soil, or bedrock, it may flow laterally through the soil (Figure 5.1). Lateral flow is common in soils on steep slopes. If water flows laterally beneath a relatively impermeable layer for some distance to an area where it is no longer confined by an overlying layer of lower permeability, it may flow upward in the soil. Lateral or upward flowing water that reaches the ground surface forms a seep or a spring. Water flowing upward to a spring, where the water pressure is greater than atmospheric pressure, is called *artesian* water.

When evapotranspiration depletes soil water above a water table, capillary forces draw water upward to replenish the capillary water. The height to which water rises above a water table ranges from no more than a few centimeters in coarse-textured soils to more than a meter in silty soils. Although capillary water potential gradients are greater in clayey soils, the conductivities are so slow that evapotranspiration keeps the water from rising as high as in silty soils, in which the rate of rise is much greater even though the potential gradients are lower. The zone of wetting above a groundwater table is called the *capillary fringe*.

5.4.4 Water Tables

The permanently saturated area belowground is called a *groundwater table*. A saturated area above an impermeable layer that is not continuous with the groundwater table, and is commonly ephemeral, is called a *perched water table*.

5.5 Evaporation and Transpiration

Much of the radiant energy from the sun that reaches the ground and some soil heat are used to vaporize water. This process is called *evaporation*. If the vaporization of water occurs in plant leaves and the vapor is transpired through stomata, it is called *transpiration*. The combination of these two processes is called *evapotranspiration*.

Most water lost from soils is through transpiration. It is largely dependent on solar radiation, leaf surface area, soil water availability, and root distribution. Because roots are generally more concentrated in surface soils and diminish with depth, soils dry more rapidly near the surface and may remain moist for many weeks into a drought season at the lower limit of root occupancy where roots are sparse.

Water loss by evaporation from soils occurs from the surface. Heat is required to vaporize water, and to produce pressure gradients to move saturated air from soils into the atmosphere. This process of transfer of soil water to the atmosphere is very slow when there is a mulch of plant detritus that retards mixing of air at the soil-atmosphere interface. Because evaporation diminishes drastically with depth, soil water depletion below the surface few centimeters depends on capillary rise to replace water evaporated in the surface soil.

5.6 Composition of Air in the Soil Atmosphere

The troposphere, which is the lower atmosphere, up to about 11 km, is mostly nitrogen, oxygen, and gases of inert elements (Table 5.4). The amounts of dust and other aerosols are highly variable. Sea salt is a common coastal aerosol, and sulfates are common aerosols in industrial areas.

Soil air is dominated by nitrogen (N_2), at nearly the same concentration as in the troposphere, but O_2 and CO_2 concentrations vary over wide ranges. Other than near the ground surface in very dry soils, the relative humidity is practically 100%. Soil organisms use O_2, lowering its concentration, and CO_2 is a product of their respiration. Microbes are concentrated around plant roots, greatly increasing the amounts of CO_2 near roots. The concentration of CO_2 increases with depth, because the movement of air is very slow in soils. It ranges from about 0.04% in the aboveground atmosphere (Chapter 11) to up to 3% in drained soils and 10% in very poorly drained soils. More air moves through soils by diffusion than by mass flow; thus, Darcy's law for mass flow is less pertinent than Fick's law for diffusion. The diffusion

Table 5.4 Composition by Volume of Gases in the Troposphere

Constant Gases		Variable Gases and Aerosols	
Gas	Percent	Gas or Particles	Parts/Million
N_2, nitrogen	78.08	CO_2, carbon dioxide	380
O_2, oxygen	20.95	CH_4, methane	1.7
Ar, argon	0.93	N_2O, nitrous oxide	0.3
Ne, neon	0.0018	O_3, ozone	0.04
He, helium	0.0005	Particles (dust, soot, etc.)	0.01–0.15
H_2, hydrogen	0.00006	CFCs, chlorofluorocarbons	0.0002

Source: Data from Ahrens, C.D., *Essentials of Meteorology*, Thomson, Belmont, CA, 2008.
Water is highly variable and ranges up to about 4% in the troposphere.
Ozone is 5 to 12 ppm in the stratosphere.

equation for the rate of a gas (for example, O_2 or CO_2) movement across a plane perpendicular to the direction of diffusion is

$$q_d = -D \, (\delta C/\delta x), \text{ Fick's law}$$

where D is the diffusivity through a soil or other medium and $\delta C/\delta x$ is the concentration gradient of the gas in the atmosphere of the air-filled pores. The rate of diffusion is dependent on the volume of air-filled pores and the connectivity and tortuosity of the pores spaces. The diffusivities of O_2 and CO_2 are about 10,000 times greater in air than in water. Consequently, gases move very slowly in wet soils in which most of the pores are filled with water; and with microbial activity, the O_2 commonly becomes so depleted in wet soils that only anaerobic organisms can survive.

5.7 Landscape Perspective

Most precipitation that falls on the ground falls through plant canopies, or flows down stems or tree trunks to the ground, and infiltrates into soils. The proportion that flows overland to streams ranges from negligible in dense forest to substantial in deserts with sparse plant cover and negligible plant litter, particularly where the soils are clayey. Crusts form on some desert and some cultivated soils, impeding infiltration of water. Water that infiltrates into soils flows downward through relatively permeable surface layers until it reaches a less permeable layer where part, or all, or the water flows laterally over the top of the less permeable layer, or until the water reaches a groundwater table. Groundwater flows laterally, unless it is trapped in a topographic low, and reaches the ground surface in springs, seeps, streams, and lakes. Water may move upward from a groundwater table where the water is confined beneath an impermeable layer that allows pressure to accumulate to greater than atmospheric pressure at the surface of the groundwater table. Artesian springs are common in limestone terrain where water flowing through cavities in bedrock to confined pools within limestone accumulates enough pressure to push water vertically to the ground surface. Less commonly, artesian pressures accumulate in soils, beneath hardpans. The source of artesian water must be higher than the outlet, because the pressure results from elevation differences.

The water budget in freely drained soils is a balance between precipitation, water intercepted by plants aboveground, drainage completely through soils, and evapotranspiration. To compute a soil water balance, surplus water, where precipitation is greater than the storage capacity of the soil, is assumed to drain from the soil, and evapotranspiration is estimated from atmospheric conditions and the estimated amount of plant available water in the soil.

Many equations have been developed to predict evapotranspiration. The more sophisticated ones are based on relative humidity, wind, solar radiation, soil moisture, and the kinds and conditions of plants that are transpiring (Geiger et al. 2003). They require much data that are commonly not available and data that are not applicable over large areas. One of the most successful schemes for predicting evapotranspiration from monthly data has been that of Thornthwaite (1948), based on latitude, AWC of the soil, and precipitation (Figure 5.4a). It makes no allowances for slope aspect and cloudiness. In order to compensate for these factors, an equation of Hargreaves presented by Jensen et al. (1990) was modified to predict potential evapotranspiration (PET) (mm/month) comparable to that of the Thornthwaite method (Thornthwaite and Mather 1955) for a station in the Klamath Mountains:

$$PET = 0.28 \text{ (mean max. } T°C - \text{mean min. } T°C)^{0.5}$$

$$(\text{mean } T°C + 1.0) \text{ (mean daily radiation, } MJ/m^2)$$

where the difference between monthly mean maximum and mean minimum temperatures is assumed to be a compensation for less readily available data for cloud cover, and potential solar radiation (MJ/m^2 per day) on different slopes at different latitudes can be computed or read from tables.

Actual evapotranspiration (AET) is dependent upon the potential evapotranspiration (PET) and the amounts of precipitation and available water (AW) in the soil. For months when the PET is greater than the precipitation, water is drawn from soils by evapotranspiration at rates that decline as the available water becomes depleted and held at greater negative potentials. For example, it might be assumed that water is lost from a soil in proportion to the 0.9th power of the ratio of available water in the soil (AW) to available water capacity (AWC) of the soil, that is, $(AW/AWC)^{0.9}$ (see Problem 5.6). Different exponents of the AW/AWC factor might be used for different types of vegetation or soil cover that have different evapotranspiration characteristics.

Walter (1974) proposed the plotting of both mean monthly temperature and precipitation on the same diagram, with the assumption that the most common rate of evapotranspiration is 2 mm per month for each degree Celsius (Figure 5.4b). His diagrams express the gross differences among different climates, but they are not appropriate for estimating the water balance at a specific site.

5.8 Global Patterns of Precipitation

Evaporation and precipitation are great at the equator and low at the poles where the incident solar radiation is minimal (Figure 5.5). Evaporation is even greater at about 20 to 30°N and S latitudes, which are called the horse latitudes or subtropical highs, than at the equator, because cloudiness at the equator reduces the amount of solar radiation that reaches the ground, or ocean, level. Westerly winds from the temperate zones north and south of the horse latitudes are relatively dry when they approach the continents, because they blow across cold ocean currents such as the Humboldt and Benguela currents that flow from higher latitudes where the oceans are cold toward the tropical zone, while easterly winds from the subtropics blow across warmer water and accumulate more moisture. Consequently, the driest areas are on the west sides of the continents (Figure 5.6). Examples of the driest areas on the west sides of the continents are the Sahara Desert in Africa and the Sonoran Desert in North America that are in the northern hemisphere and the Kalahari Desert in Africa and the Atacama Desert in South America that are in the southern hemisphere, as are the deserts of Western Australia. Some deserts are in the rain shadows of mountains and do not conform to this pattern. Rain shadow arid areas, or deserts, such as the Gobi Desert in Central Asia and the Great Basin Desert in North America are not represented in Figure 5.6.

Global climatic warming will increase the total amount of precipitation on Earth as increased temperatures increase evapotranspiration, but as the precipitation increases in some places, it will decrease in others. The patterns of changing precipitation are very difficult to predict.

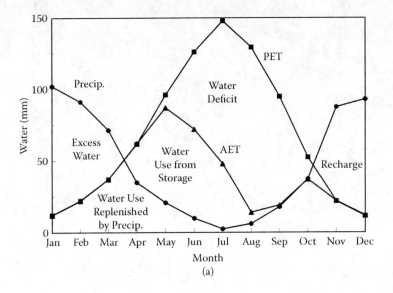

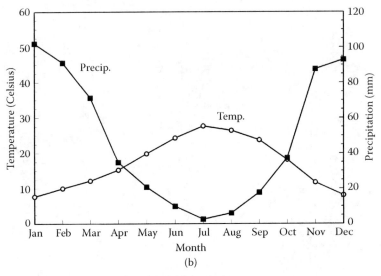

Figure 5.4 Annual precipitation and water use, or loss, patterns at Red Bluff, California. (a) Mean monthly precipitation and potential and actual evapotranspiration, assuming a soil AWC of 200 mm. (b) Monthly mean temperatures and precipitation at Red Bluff, with a scale of 2 mm of precipitation for each degree Celsius.

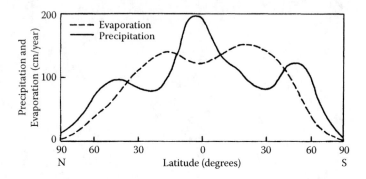

Figure 5.5   Mean annual precipitation and evaporation by latitude from the North Pole across the equator to the South Pole. (Adapted from Henderson-Sellers, A., and P.J. Robinson, *Contemporary Climatology*, Longman, Essex, England, 1986.)

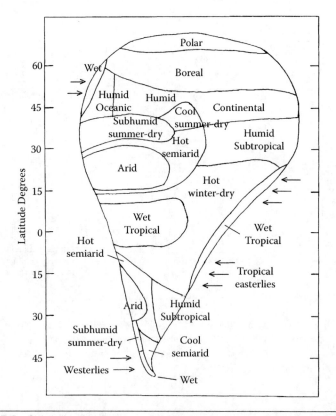

Figure 5.6   Global climatic patterns: a schematic presentation. (Based on maps and diagrams from Trewartha, G.T., *An Introduction to Climate*, McGraw-Hill, New York, 1968; McKnight, T.L., *Physical Geography: A Landscape Appreciation*, Prentice-Hall, Englewood Cliffs, NJ, 1984; and Strahler, A.N., and A.H. Stahler, *Modern Physical Geography*, Wiley, New York, 1987.)

Problems

5.1. A loamy sand soil has a gravimetric AWC (water mass/soil mass) of 6% with < 1% organic matter, and a soil of the same texture has an AWC of 10% with 3% organic mater. The bulk densities are 1.5 Mg/m^3 for the soil with low organic matter and 1.1 Mg/m^3 for the soil with 3% organic matter. What are the volumetric water contents (water volume/soil volume) of the soils?

Solution:
Soil with < 1% OM: Soil water (Mg/m^3) = 1.5 Mg/m^3 × 0.06 Mg water/Mg soil = 0.09

Soil with 3% OM: Water (Mg/m^3) = 1.1 Mg/m^3 × 0.10 Mg water/Mg soil = 0.11

A milligram of water is a cubic meter; therefore, the answers are equal to 0.09 and 0.11 m^3/m^3, or 9 and 11%. The difference is only 2%, compared to a 4% difference for the gravimetric contents.

5.2. A loam that has a gravimetric AWC (water mass/soil mass) of 14% with < 1% organic matter and a soil of the same texture has an AWC of 18% with 3% organic mater. The bulk densities are 1.5 Mg/m^3 for the soil with low organic matter and 1.1 Mg/m^3 for the soil with 3% organic matter. What are the volumetric water contents (water volume/soil volume) of the soils?

Solution:
Soil with < 1% OM: Soil water (Mg/m^3) = 1.5 Mg/m^3 × 0.14 Mg water/Mg soil = 0.21
Soil with 3% OM: Water (Mg/m^3) = 1.1 Mg/m^3 × 0.18 Mg water/Mg soil = 0.20

A milligram of water is a cubic meter; therefore, the answers are equal to 0.21 and 0.20 m^3/m^3, or 21 and 20%. The 4% difference in the gravimetric water contents is practically nullified when the water contents are presented on a volumetric basis.

5.3. Given a soil with a loam surface that is 20 cm thick, a subsoil that is a gravelly clay loam with 25% (volumetric) gravel, and bedrock at 60 cm depth, how much plant available water will the soil hold?

Solution: Multiply the thickness of each layer by its AWC (Table 5.1), divide by 100, and sum the results for all layers. The AWC of the layer with stones is reduced by the volume of the stones.

Surface soil: 20 cm × 21/100 = 4.2 cm

Subsoil: (60 cm – 20 cm) × 19/100 × (100 – 25)/100 = 5.7 cm

Soil, total: AWC = 4.2 cm + 5.7 cm = 9.9 cm

5.4. Absent any drainage restrictions, how rapidly will water flow from a completely saturated soil with a 15 cm sandy loam surface (K_{sat} = 3.6 cm/h) and a clay loam subsoil 60 cm thick (K_{sat} = 0.01 cm/h)?

Solution: The hydraulic head is equivalent to the depth of the bottom of the more restrictive layer, the subsoil, assuming that precipitation or melting snow keeps the soil above the more restrictive layer saturated, and the length of the critical flow path is the thickness of the more restrictive layer.

Thus, $q = K_{sat} (\Delta H/L) = 0.01$ cm/h (75 cm/60 cm) = 0.0125 cm/h. The rate will decrease to 0.01 cm/h as the surface soil drains to field capacity and water continues to drain from the subsoil.

5.5 With a mean potential incident solar radiation of 170 W/m² over 12 h/day, a mean monthly (June) temperature of 19.0°C, a mean maximum of 29.7, and a mean minimum of 8.6°C, what is the monthly potential evapotranspiration (use the PET equation)?

Solution:

PET (mm/month) = 0.28 (mean max. T°C – mean min. T°C)$^{0.5}$ (mean T°C + 1.0) (daily MJ/m²)

Remember that W = J/s.

Daily irradiation = 170 W/m² × (MJ/10⁶ J) × (12 h × 3600 s/h) = 7.344 MJ/m²

PET = 0.28 (29.7 – 8.6)$^{0.5}$ (19.0 + 1.0) × 7.344 = 189 mm/month

5.6. At a site with July precipitation of 13 mm, July potential evapotranspiration (ET) of 53 mm, soil with antecedent (end of June) water content of 115 mm, and soil AWC of 120 mm, how much water might be lost from the soil in July, and how much AW would remain stored in the soil? If in August the precipitation is 32 mm and the potential ET is 49 mm, how much water will be in the soil at the end of August?

Solution: Assume evapotranspiration loss from the soil at a rate proportional to $(AW/AWC)^{0.9}$, as in Section 5.7. Because the 0.9 exponent is not a precise number, results are rounded to the nearest millimeter.

July: ET from precipitation = 13 mm, ET from soil = $(53 - 13) (115/120)^{0.9}$ = 38 mm lost from soil
AW stored in the soil at end of July = 115 − 38 = 77 mm
August: ET from precipitation = 32 mm, ET from soil = $(49 - 32) (77/120)^{0.9}$ = 11 mm loss
AW stored in the soil at end of August = 77 − 38 = 39 mm

Questions

1. Are soils sufficiently homogeneous that precipitation or irrigation water will descend downward in a uniform wetting front?
2. Smectite clay has much greater surface area than kaolinite clay (Table 2.4) but retains hardly any more water at the wilting point (−1.5 MPa) of plants. Why does smectite *not* hold much more water than kaolinite at this pressure?
3. What might cause a soil wetting front from rainwater on a dry soil to descend through the soil irregularly, rather than advancing uniformly at the same rate over areas of a square meter or more?
4. Is the force in Darcy's law a gravitational force or a matric force?
5. Is the height that a capillary fringe rises above a groundwater table generally greater in clayey or silty soils?
6. Evapotranspiration from well-vegetated soils with deeply rooted plants is generally much greater than evaporation from adjacent soils lacking plant cover. Why?
7. How does wind affect evapotranspiration rates?

Supplemental Reading

Ahrens, C.D. 2008. *Essentials of Meteorology.* Thomson, Belmont, CA.

Bonan, G. 2008. *Ecological Climatology.* Cambridge University Press, New York.

Geiger, R., R.H. Aron, and P. Todhunter. 2003. *The Climate Near the Ground.* Rowan and Littlefield, Lanham, MD.

Hillel, D. 1980. *Fundamentals of Soil Physics.* Academic Press, Orlando, FL.

Soil Survey Division Staff. 1993. *Soil Survey Manual.* U.S. Department of Agriculture Handbook 18. U.S. Government Printing Office, Washington, DC.

Soil Survey Staff. 1993. *National Soil Survey Handbook.* Title 430-VI. U.S. Government Printing Office, Washington, DC.

Taylor, S.A., and G.L. Ashcroft. 1972. *Physical Edaphology.* Freeman, San Francisco.

6

SOIL CLASSIFICATION—
KINDS OF SOILS

There are billions of pedons on Earth. We cannot investigate all of them, but we can group pedons into a manageable number of classes and investigate pedons representative of soils in each class. Then, we can make predictions about similar soils that have not been investigated intensively, but can be assigned to known classes of soils. Thus, soil classification is more than an academic exercise—it allows us to plan land resource allocation and management more effectively utilizing a vast store of knowledge about soils and related land capability and to predict the responses of different soils to management.

The first comprehensive system of soil classification was developed by Russians in the latter half of the nineteenth century, following the example of Vasily Dokuchaev. A basic feature of the Russian system was the concept of azonal, zonal, and intrazonal soils. Soils characteristic of specific climatic zones and biomes were called zonal soils; for example, typical soils in cold conifer forests were called Podsols, and black soils of grasslands were called Chernozems. Soils lacking development were called azonal soils, and soils with development dependent on special conditions, conditions other than good drainage or parent materials other than those of the zonal soils, were called intrazonal soils. A major problem with the zonal concept is in deciding what soils are zonal. Within the same climatic zone, soils with different parent materials may differ greatly. Which one should be designated the zonal soil?

Even though the original Russian classification system was highly subjective, modifications of the Russian system were adapted for application in other countries. Then in the third quarter of the twentieth century, a more objective system of soil classification was developed in the United States. The effort was directed by Charles Kellogg of the U.S. Department of Agriculture (USDA). He chose Guy Smith to lead development of the new system. It went through several drafts, called the seven approximations, that were tested by application in the USDA and reviewed internationally. A complete edition of *Soil Taxonomy* was published in 1975 (Soil Survey Staff 1975), but revision has continued with the publication of a revised *Keys to Soil Taxonomy* every few years and a second publication of *Soil Taxonomy* in 1999.

Even though *Soil Taxonomy* has objective keys, it has involved many arbitrary choices and the classes are artificial. Attempts are generally made to make the classes more natural by choosing class limits that are less arbitrary. For example, if many soils had less than 30% clay, many had 34 to 40% clay, and few had 31 to 33% clay, it would be more natural to set a class limit at about 32% clay than at 35% clay.

Although many countries have developed soil modern classification systems, only two systems are in multinational use. These two are the *Soil Taxonomy* of the U.S. Department of Agriculture (Soil Survey Staff 1999) and a somewhat similar system initiated in the Food and Agriculture Organization (FAO) of the United Nations and currently identified as the World Reference Base (WRB) for Soil Resources (IUSS Working Group 2006). Both systems are hierarchical, but the hierarchy has been less completely developed in the WRB. The USDA system is complete, although still being revised, and is the one covered in more detail here.

6.1 Organization of the USDA System, *Soil Taxonomy*

The hierarchy of *Soil Taxonomy* contains six levels, or categories: order, suborder, great group (or group), subgroup, family, and series. Phases of soil series are used in mapping, but those phases are not a part of *Soil Taxonomy*. In an older USDA system, before *Soil Taxonomy*, soil series and families were established by mapping and higher categories were more hypothetical, based on concepts of soil development. By building the system up from soil series to families, and extending higher categories from the top down to subgroups, it was difficult to join soil families in subgroups. Although soil series are still based on mapping, *Soil Taxonomy* is organized from the top down. However, no classes are set up in higher categories unless they are represented by mapped soil series at the bottom of the hierarchy. The join between families and series is still unsatisfactory, because the ranges of soil properties within a series are made to fit artificially within the limits of a family, even though those limits may not be the ones that seem natural in mapping the soils.

Currently there are 12 orders, dozens of suborders, hundreds of groups, thousands of subgroups, and probably tens of thousands of families and hundreds of thousands of soil series. No one knows how many soil series have been mapped on Earth.

6.2 Diagnostic Horizons

Diagnostic horizons are fundamental to *Soil Taxonomy*. They are precisely defined layers, whereas the horizons of pedon description, other than O horizons, are not defined quantitatively. A diagnostic horizon incorporates one or more horizons recognized in a pedon description. Diagnostic horizons occurring in surface soils are called epipedons. Epipedons and diagnostic subsurface horizons are listed in Table 6.1. Complete definitions of some of the diagnostic horizons are very complex, and those details need not concern us here. More important is an appreciation of the processes involved in developing the diagnostic horizons.

6.2.1 Epipedons

Root concentrations and animal activities are greatest in surface soils. Animals facilitate the decomposition of organic matter, and some of them carry organic matter from aboveground plant detritus down into the soil. Root decay and incorporation of organic matter into soils by animals generally add enough organic matter to surfaces soils to make them dark greyish brown to black.

Six kinds of surface horizons are recognized based largely on their color, concentration of organic matter, and thickness. An organic surface horizon 20 to 40 cm thick resting on mineral soil is called a *folic* epipedon if it is well drained or a *histic* epipedon if it is saturated with water during the growing season. It can be thicker, up to 60 cm, if the organic matter is no more than slightly decomposed. A thick (thickness > 30 cm) black horizon with concentrated organic matter (organic carbon > 6%) and andic soil properties is called a *melanic* epipedon. Other inorganic, or mineral, surface layers are *mollic* or

Table 6.1   Diagnostic Horizons

Horizon	Explanatory Phrase[a]
Surface Layers, or Epipedons	
Anthropic	Evidence of human disturbance or high acid-soluble P_2O_5
Folic	Thick layer of well-drained organic material, generally 20 to 40 cm thick
Histic	Thick layer of poorly drained organic material, generally 20 to 40 cm thick
Melanic	Thick black layer with much organic matter and andic soil properties
Mollic	Thick dark-colored layer, base saturation > 50%, general thickness > 25 cm
Ochric	Lighter-colored or thin dark-colored layer that lacks the diagnostic features of other epipedons
Plaggen	Centuries-old accumulation of manure > 50 cm thick
Umbric	Thick dark-colored layer, base saturation < 50%, general thickness > 25 cm
Subsurface Horizons, or Layers	
Albic	Light-colored eluvial horizon
Argillic	Illuvial accumulation of clay
Calcic	Illuvial accumulation of lime, or calcium carbonate
Cambic	Evidence of incipient soil development, such as higher chroma or soil structure
Duripan	Layer cemented by illuvial accumulation of silica
Gypsic	Illuvial layer of secondary gypsum accumulation
Kandic	Similar to an argillic horizon, but with low CEC
Ortstein	Layer of cemented iron compounds or spodic materials
Oxic	Lack of weatherable minerals and low CEC, lacking the illuvial clay
Petrocalcic	Horizon cemented by illuvial accumulation of calcium carbonate
Placic	Thin hardpan cemented by iron and commonly manganese compounds
Salic	Accumulation of salts more soluble than gypsum
Spodic	Accumulation of aluminum and iron organic complexes with highly variable charge (CEC varies greatly with pH), generally beneath an albic E horizon
Sulfuric	Mineral or organic material with pH < 3.5 and containing sulfates

[a]   This is a selective list with abbreviated definitions. Complete definitions are commonly complex (Soil Survey Staff, 1999). Also, anhydrous and aquic conditions are defined precisely in *Soil Taxonomy* utilized in soil classification.

umbric epipedons if the dark-colored surface is thicker than 25 cm. Thinner mollic and umbric epipedons occur in shallow soils, but the thickness requirements are too complex to explain in one or two sentences (Soil Survey Staff 1999). The base saturation, based on a cation-exchange capacity (CEC) at pH 7, is > 50% in mollic epipedons and < 50% in umbric epipedons—no other distinctions are made between them. Mineral surface layers that do not meet the requirements of mollic or umbric epipedons are called *ochric* epipedons.

A surface or near-surface layer bleached by leaching of iron is an *albic horizon*, rather than an *epipedon*. Albic horizons generally occur beneath Oa or A horizons. Where developed between high and low levels of a fluctuating water table, an albic horizon may be a subsurface layer.

Human disturbance is readily apparent in some soils, and anthropic and plaggen epipedons are recognized to acknowledge that disturbance. Plaggen epipedons are the result of animal bedding or manure addition for decades or centuries, and they are inextensive beyond Western Europe. The anthropic epipedon definition has not been completely developed yet. Plow layers are artificially disturbed layers that are not generally considered anthropic horizons unless they have high concentrations of phosphorus (P_2O_5 > 250 Mg/kg) that are associated with human activity.

6.2.2 Subsurface Diagnostic Horizons, or Layers

There are many kinds of subsurface diagnostic horizons, resulting from many soil processes acting on different soil parent materials. These soil processes involve weathering, leaching, local transport of materials, accumulation of materials, and shrink-swell.

Weak expression of any of the soil processes is evidence for a *cambic* horizon. This evidence may be increased chroma (redder color) resulting from weathering that releases iron, which then concentrates in spots or on surfaces and is oxidized, loss of carbonates by leaching, or development of blocky or prismatic structure by repeated shrinkage and swelling of soil. The expression, or *grade*, of a blocky or prismatic structure may be increased by accumulation of illuvial clay on ped faces. When the increase of clay is 20% or more of the amount initially present in the parent material, and at least some of the clay is illuvial, the layer is called an *argillic* horizon. Two variations from the argillic horizon are recognized as diagnostic horizons: an argillic horizon in which sodium has accumulated is called a *natric* horizon, and one in which weathering has reduced the CEC (pH 7) of clay to < 16 me/100 g is called a *kandic* horizon.

An extremely weathered layer with sparse evidence of illuvial clay is called an *oxic* horizon. The clay minerals in oxic horizons are predominantly kaolinite, gibbsite, goethite, and hematite. These are all low-activity clays lacking the basic cations present in Mg-chlorites, hydrous micas, and smectites that are common in less weathered soils.

The majority of subsurface diagnostic horizons are recognized by accumulations of materials, mostly illuvial accumulations. These horizons and their definitive materials include *calcic* horizons with calcium carbonate, *gypsic* horizons with gypsum, *natric* horizons with clay and sodium, *salic* horizons with salts more soluble than gypsum, and *spodic* horizons with aluminum, iron, and organic matter. Where these horizons are cemented they are *petrocalcic* horizons with calcium carbonate or calcite, *duripans* with silica in opal or chalcedony (cryptocrystalline quartz), *ortstein* with iron or iron-organic complexes, and *placic* horizons with iron-organic complexes. Placic horizons are thinner (thickness < 25 mm) than ortstein layers, and the surfaces of placic horizons are generally wavy or convoluted. A subsurface root-restricting layer that is hard or very hard when dry, brittle when moist, and slakes in water is called a *fragipan*.

6.3 Soil Temperature and Moisture Regimes

Climate is a major factor in soil development, plant community dynamics, and soil management. Because soils are classified by soil properties, rather than by external factors such as aboveground climate, soil climate is utilized in *Soil Taxonomy*. Soil temperature and moisture regimes have been establish to represent soil climate. Reliable ascertainment of these regimes requires data collected over several years, because differences from one year to another are commonly great.

6.3.1 Soil Temperature Regimes

Differentiation of soil temperature regimes is based primarily on mean annual temperature at 50 cm depth, or at the soil-bedrock contact in shallow soils (Table 6.2). The regimes, from equator to poles, or high mountains, are *hyperthermic* or hot tropical, *thermic* or warm subtropical, *mesic* or cool temperate, *frigid* or cold temperate, *cryic* or very cold boreal (or very cold austral), and *pergelic* or frozen polar. The cryic regime is differentiated from the frigid regime by being colder during summer. Regimes where the means of the coldest and warmest months differ by $T < 5°C$ are given *iso-* prefixes: isohyperthermic, isothermic, isomesic, and isofrigid.

Table 6.2 Soil Temperature Regimes (STRs)

	Soil Temperature at 50 cm Depth		
	Mean Annual Temperature	Annual Range[a]	
		ΔT > 5°C	ΔT < 5°C
Subjective Term	°C	Soil Temperature Regime	
Hot	T > 22	Hyperthermic	Isohyperthermic
Warm	15–22	Thermic	Isothermic
Cool	8–15	Mesic	Isomesic
Cold	0–8	Frigid	Isofrigid
Very cold	0–8	Cryic[b]	—
Frozen	T < 0	Pergelic	—

[a] Iso regimes occur in soils with <5°C between means for the coldest and warmest months.

[b] Soils in cryic regimes are never warmer than 8°C, even in summer.

6.3.2 Soil Moisture Regimes

Soil moisture regimes are the primary differentiating features at the order level in one order (Aridisols), the great group level in two orders, and the suborder level in eight orders. This reflects the great importance of soil moisture in soil development, plant distribution and productivity, and soil management.

Soil moisture regimes are based on the moisture status of the moisture control section during growing seasons, or when the soil temperature is 5°C or greater at 50 cm depth. The soil moisture control section is the depth of wetting by 2.5 to 7.5 cm of water, which is between 10 and 30 cm in silty soils to about 50 to 150 cm or deeper in stony sandy soils. The moisture regimes are very dry, *arid*; dry, *ustic*; summer dry, *xeric*; moist, *udic*; always moist, *perudic*; and wet, *aquic* regimes (Table 6.3). Aquic conditions as interpreted by redoximorphic features are utilized in defining soil classes, rather than the aquic soil moisture regime. The perudic regime is not utilized in defining soil classes.

Between the aridic regime of deserts and the udic regime of humid regions are the ustic and xeric regimes of semiarid to subhumid regions. Ustic soils are moist long enough during a growing season to produce many kinds of crops without irrigation. Xeric soils, which occur in Mediterranean climates, are too dry following the summer solstice to grow more than a few kinds of crops, mostly small grains, without irrigation. Soils with cryic soil temperature regimes cannot

Table 6.3   Soil Moisture Regimes (SMRs)

Moisture Regime	Definitive Characteristics in Moisture Control Section[a]
Aquic	Saturated with water in growing season long enough to produce redoximorphic features
Aridic	Dry throughout for more than one-half of growing season and no consecutive 90 days in growing season when soils are moist in some parts
Udic	Not dry in any part for any 90 days in growing season
Ustic	Dry in some parts for any 90 days in growing season
Xeric	Dry throughout for at least 45 consecutive days in 4 months following summer solstice

[a] Depth of the soil moisture control section is about 10 to 30 cm in fine-silty soils and ranges to 50 to 100 cm, or more, in sandy soils that hold only 5%, or less, plant available water.

have either ustic or xeric soil moisture regimes, because the growing season is too short (less than 90 days). And xeric soils, which are moist for at least 1.5 months when soils are cold following the winter solstice, are not recognized in hyperthermic soil temperature regimes.

6.4   Order and Lower Levels of *Soil Taxonomy*

A person can become familiar with all taxa at the order level, because there are only 12 orders (Table 6.4). The number of taxa increases so markedly with each step down the hierarchy that it is not practical to attempt familiarity with every suborder, and it is practically impossible to become familiar with every subgroup. However, many pedologists are familiar with, or cognizant of, most of the diagnostic soil horizons

Table 6.4   Orders of *Soil Taxonomy* and Properties or Characteristics that Define Them

Order	Formative Ending	Definitive Characteristics
Alfisols	Alf	Argillic horizon with base saturation > 35% at pH 8.2
Andisols	And	Andic properties
Aridisols	Id	Soil climate (moisture) and soil development
Entisols	Ent	Lack of soil development
Gelisols	Gel	Permafrost
Histosols	Ist	Organic material > 40 or 60 cm thickness
Inceptisols	Ept	Incipient soil development
Mollisols	Oll	Mollic epipedon and high base saturation (base saturation > 50% at pH 7)
Oxisols	Ox	Highly weathered soil without argillic horizon
Spodosols	Od	Spodic horizon
Ultisols	Ult	Argillic horizon with base saturation < 35% at pH 8.2
Vertisols	Ert	Mixing by shrink-swell

and other features that are utilized to differentiate and characterize the classes. These features are so closely linked to the taxonomic names that an experienced pedologist who sees or hears the name of a subgroup or family that is unfamiliar to him or her can readily produce a somewhat accurate mental picture of the soil and infer many of its properties and characteristics. There is no need to remember every class of soil to know something about each.

6.4.1 *Orders of* Soil Taxonomy

The orders of *Soil Taxonomy* are designed to reflect processes of soil development. Only the Histosols are restricted to a single soil parent material, organic material, although most Andisols develop in volcanic parent materials. Climate, or soil climate, appears only in the definitions of Aridisols and Gelisols at the order level, but soil temperature and moisture are very important at the suborder and lower levels in all orders of *Soil Taxonomy*.

Soils in initial phases of soil development, lacking diagnostic horizons other than an ochric epipedon, are Entisols. Vertisols may lack diagnostic epipedons too, but because they are mixed by shrinkage and swelling, rather than because they are in initial phases of soil development. Soil development in Entisols may lead directly to Aridisols, Mollisols, Andisols, Spodosols, or Inceptisols with mollic, umbric, melanic, or histic epipedons or cambic, calcic, gypsic, salic, or placic horizons, or fragipans. Inceptisols are generally intermediate in a succession from Entisols to Alfisols or Ultisols with argillic horizons. Aridisols and Mollisols may have argillic horizons too, and many other kinds of diagnostic horizons. Both Aridisols and Gelisols may have many different kinds of diagnostic horizons, because they are defined by dry or freezing climates, rather than by soil development, although most Aridisols have some pedological development. There are many paths in the succession, from Entisols lacking soil development to Oxisols with oxic horizons. Soil development will be discussed more thoroughly in Chapter 7.

6.4.2 *Suborders, Groups, and Subgroups*

Soils in each subgroup have the properties required of the great group, suborder, and order above in the hierarchy of *Soil Taxonomy*. Naming a taxon consists of beginning with an order ending (Table 6.4), adding on the left formative elements for a suborder and a great group, and attaching an adjective for a subgroup.

As an example, consider a cold soil with an argillic horizon, udic soil temperature regime, very high basic cation status, and ochric epipedon containing volcanic ash. We find in the key to orders (Soil Survey Staff 1999) that this soil is an Alfisol with a formative ending *alf*. In the key to Alfisols, we find that the soil is a Boralf (formative element *bor* + *alf*), because it is cold and moist during summer; in the Eutroboralf group, because it has very high basic cation status; and in a Vitrandic Eutroboralf subgroup, because it contains volcanic glass in the surface soil. If we had only the name, we could infer everything about the soil that we needed to know to classify it. It must have an argillic horizon, because it is an Alfisol, etc. Thus, there is considerable information about a soil in its taxonomic name, and we do not have to have previous knowledge about a particular soil to infer what it is like. We can infer, without guessing, that the Vitrandic Eutroboralf is cold and moist through most of the summer, has a subsoil with illuvial clay (argillic horizon) of basic cation status, and a surface horizon containing pyroclastic materials and some amorphous aluminum and iron.

6.4.3 Soil Families and Series

The family level is the most complex in *Soil Taxonomy*. A family name consists of the subgroup name plus terms for a textural class, a mineralogical class, and a soil temperature regime. A clay activity class is added for soils with mixed mineralogy classes. For example, a Vitrandic Eutroboralf might be in a fine-loamy, mixed, superactive, frigid family of that subgroup. A frigid family class is implied by the soil being in a Eutroboralf subgroup. Therefore, the entire family name would be a fine-loamy, mixed, superactive Vitrandic Eutroboralf. Fine-loamy is a textural class with 18 to 35% clay, silt > 15%, and less than 35% pebbles and cobbles; mixed is a mineralogy class in which the silt and sand fractions are not dominated by any one kind of mineral; and superactive is an activity class in which clay is very active (CEC > 60 me/100 g of clay). Some soils have soil reaction (pH) classes in addition to other kinds of classes. All of the detail about families is in *Keys to Soil Taxonomy* if you want to learn more about the designation of soil families.

All families contain at least one soil series, and some families have several. Families are most commonly differentiated into soil series by soil depth class, but there are many other soil features that are utilized to differentiate series within a family. Any soil property or characteristic that might be important in soil use and management can be utilized to separate soil series. Soil series are given local names from locations where they are found; for example, the Miami series may be mapped near the Miami River. Fortunately, series names stand alone, rather than being added to the commonly very long family names.

In mapping, slope phases and commonly other phases, such as erosion, are added to a series. These phases are not considered to be part of *Soil Taxonomy*. Many phases indicate landscape features that are not strictly soil properties or characteristics. They are mapped for utilization in land use planning and management.

6.5 World Reference Base (WRB)

The WRB is based on the same principles as the U.S. *Soil Taxonomy*, but without the soil temperature and moisture regimes. Instead of the 12 orders of the U.S. *Soil Taxonomy*, the WRB has 30 soil groups. The soil group names are Histosols, Cryosols, Anthrosols, Leptosols, Vertisols, Fluvisols, Solonchaks, Gleysols, Andosols, Podzols, Plinthosols, Ferralsols, Solonetz, Planosols, Chernozems, Kastanozems, Phaeozems, Gypsisols, Durisols, Calcisols, Albeluvisols, Alisols, Nitisols, Acrisols, Luvisols, Lixisols, Umbrisols, Cambisols, Arenosols, and Regosols.

Scores of adjectives have been defined to modify these 30 soil groups. A Leptic Luvisol, for example, is a Luvisol with bedrock within 100 cm depth, or more specifically, an Epileptic Livisol if the bedrock is between 50 and 100 cm depth. A second modifier can be added in parentheses; for example, an Epileptic Luvisol (chromic) if a layer > 30 cm thick in the subsoil has a hue redder than 7.5 YR in the Munsell color system.

6.6 Features of Soils in the Twelve Orders of *Soil Taxonomy*

Soils in most of the 12 orders have some relations to climate, but only the Gelisols and Aridisols are restricted to specific climates. Only the Histosols have specific parent materials, although the Andisols have a volcanic material bias. Poorly drained soils are separated from well-drained soils at lower levels in *Soil Taxonomy*, but not at the order level. A comparison of global climates and the global distributions of soils (Section 6.7) indicates the importance of climate in the development and classification of soils. Brief descriptions of the 12 orders and their distributions follow in the order that they appear in the keys to *Soil Taxonomy*.

6.6.1 Gelisols

Soils that remain frozen in at least some part of the upper meter throughout the year are Gelisols. Consequently, they are in polar regions (Figure 6.1E), and some are in very high mountains elsewhere. Some Gelisols that are too dry to contain continuous layers of ice may contain lenses and veins of ice, or only ice crystals. The vegetative cover of Gelisols is generally sparse in the drier areas, which have been called polar deserts, to mosses and lichens, with few small trees and shrubs in wetter areas. The plants on the Gelisol of Figure 6.1J are sparse black spruce (*Picea mariana*), few shrubs and forbs, common sedges and rushes, plentiful mosses, and abundant lichens (mainly reindeer lichen (*Cladina* sp.)) (Figure 6.2). The soil is in the Orthel suborder, because it is not organic (Histels) and lacks evidence of cryoturbation (Turbels). It is in the Historthel great group, because it has a histic epipedon.

6.6.2 Histosols

Histosols have organic matter parent materials. They are almost exclusively poorly drained, because the anaerobic conditions of soils saturated with water retard the decomposition of the plant detritus that is the parent material of organic soils; Folists are more well-drained exceptions. Poorly drained organic soils occur in all climatic regions. Some of the largest areas of Histosols are on large lacustrine plains and on the deltas of large rivers, but they are too small to delineate on a global scale (Figure 6.3). Histosols are most common in the Spodosol region, where the cold climate inhibits the decomposition of organic matter. The vegetative cover of Histosols is commonly grasses, sedges, and rushes, with abundant mosses in the colder areas. Some Histosols are in forests, and Histosols can be found on mangrove (*Rhizophora mangle*) tidal flats.

A major feature used to differentiate Histosols at the suborder level is the fiber content. From high to low fiber contents, the suborders are Fibrist, Hemist, and Saprist. The bulk densities of organic soils increase as decomposition reduces the fiber content. Hemists have bulk densities of about 0.1 Mg/m^3, and Saprists have bulk densities up to about 0.2 Mg/m^3 (Nichols and Boelter 1984). Histolsols have bulk densities much greater than about 0.2 Mg/m^3 only if they contain inorganic particles. The Histosol in Figure 6.1F is in the Saprist suborder, because the soil is dominated by Oa horizons of well-decomposed organic matter.

6.6.3 Spodosols

Spodosols are soils with spodic horizons, horizons that contain amorphous aluminum compounds or organic matter, or both, that have been leached downward from overlying Oa, A, or E horizons. Aluminum and iron are commonly chelated by the organic compounds that leach downward and accumulate in the spodic horizon. Spodosols are characteristic of boreal forests, but they occur in tropical areas also. Figure 6.1(J) shows a typical Spodosol that is extremely acidic above the spodic horizon and strongly acidic in the spodic horizon. The E horizon can be much thicker in tropical Spodosols.

6.6.4 Andisols

Andisols are soils with andic soil properties. Soils with andic properties contain poorly ordered aluminosilcates (allophane and imogolite) or aluminum-humus complexes, or both, and may contain hydrated iron oxide (ferrihydrite); they have either bulk densities < 0.9 Mg/kg or more than 5% volcanic glass in the coarse silt to sand fraction. Andisols are common in tephra (volcanic ejecta). They can form, less commonly, in soils with nonvolcanic parent materials in cool humid climates. Figure 6.1C shows a Vitrixerand in loess and volcanic ejecta resting on glacial till. It is in the Xerand suborder because it dries during summers, and in the Vitrixerand great group because it contains volcanic glass from the tephra. Andisols are common along mid-ocean

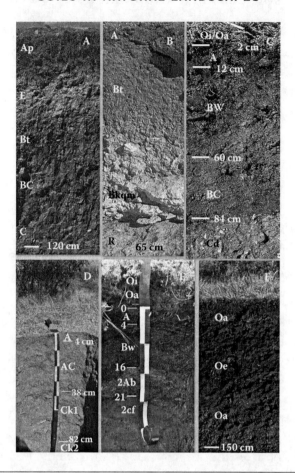

Figure 6.1   Soils of the 12 orders of *Soil Taxonomy*. (A) Alfisol in glacial till, Michigan. (B) Aridisol on basalt, Idaho. (C) Andisol in loess over glacial till, Washington. (D) Entisol on a granitic alluvial fan, California. (E) Gelisol in alluvium from granitic rocks, Alaska. (F) Histosol in a wet meadow, Idaho. (G) Inceptisol on serpentinite, Costa Rica. (H) Mollisol in glacial till (Houdek series, state soil), South Dakota. (I) Oxisol on weathered basalt (Molokai series), Hawaii. (J) Spodosol in glacial till (Chesuncook series, state soil), Maine. (K) Ultisol on metamorphosed sandstone, Arkansas. (L) Vertisol in lacustrine deposits, Idaho. (Photos D, E, and G by author and the remainder from NRCS documents.)

ridges and above mantle plumes (for example, the Hawaiian Islands and the Snake River–Yellowstone area) and on continental margins where oceanic plates have been subducted beneath them, producing volcanic activity above the sinking plates. There is a narrow strip of the volcanic activity and Andisols around the Pacific Rim from the southern tip of South America through the Aleutian Islands to the Philippines and New Zealand, and a narrow strip adjacent to the Java Trench. The distributions of active volcanic regions and most Andisols are not related

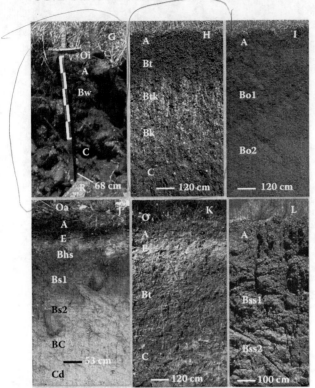

Figure 6.1 (*Continued*)

to climate, and there are none around the Atlantic Ocean (Iceland and other Atlantic volcanic islands are small or nearer the centre of the Ocean). No Andisol polygons are shown in Figure 6.3.

6.6.5 Oxisols

Oxisols are highly weathered soils, with less than 10% (< 100 g/kg) weatherable minerals and low CECs. They are produced in tropical climates (Figure 6.3). Figure 6.1I shows a Rhodic Eutrustox in Hawaii. The suborder is Ustox, because it dries sometime during the plant growing season; it is a Eutrustox, because of base saturation > 35%; and it is in the Rhodic subgroup of the Eutrustox great group, because it is dark red.

6.6.6 Vertisols

Vertisols are clayey soils in which the clay is dominated by smectites that expand and shrink upon wetting and drying. They are in climates

Figure 6.2   Landscape of a Gelisol (Figure 6.1E) on a granitic pediment. The Alaska pipeline is aboveground here; it is belowground where there is no permafrost (frozen ground). This is near the northern limit of black spruce (*Picea mariana*) in Alaska, but white spruce (*Picea glauca*) occurs further north.

with distinct dry seasons in which the clays shrink and wet seasons in which the clays expand. The shrinking and swelling displaces soil along oblique (neither vertical nor horizontal) planes (Figure 6.1L). Mixing, or homogenization, of the soils and large vertical cracks is a characteristic feature. Movement along oblique planes smears the clay to form slickensides (Figure 3.2C). Vertisols are common on lacustrine plains and coastal plains in areas where the soils have ustic or xeric soil moisture regimes. There are large areas of Vertisols in eastern Africa, India, and Australia (Figure 6.3). The Vertisol of Figure 6.1L is a Xeric Epiaquert. The suborder is an Aquert, because the soil is wet through winters; it is an Epiaquert, because it is wetted from the surface, rather than by groundwater; it is in a Xeric subgroup, because it dries during summers.

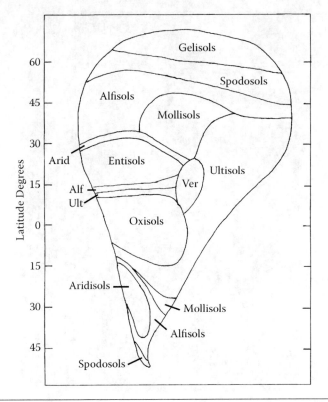

Figure 6.3   A global schematic map of soils in the 12 orders of *Soil Taxonomy*. Histosols occur in any of the soil areas, but they are most common with the Spodosols.

6.6.7 *Aridisols*

Aridisols have aridic soil moisture regimes and either cambic, argillic, natric, salic, calcic, petrocalcic, gypsic, or petrogypsic horizons, or duripans. They generally have accumulations of Ca-carbonate, which crystallizes to calcite, and many Aridisols contain gypsum ($CaSO_4 \cdot 2H_2O$). Leaching prevents these compounds from accumulating in soils of wet climates. Some soils with salic horizons have restricted drainage that causes them to be wet above a meter depth for more than a month each year. The Aridisol in Figure 6.1B is in the Durids suborder, because it has a duripan (the Bkqm horizon), and it is in the Argidurids great group, because it has an argillic horizon. The vegetation on Aridisols is most commonly cacti and other succulent plants, sparse trees, few shrubs, and ephemeral forbs or grasses (Figure 6.4).

Figure 6.4   An Aridisol, a Duric Petrargid, on a serpentized peridotite alluvial fan in Baja California Sur. The tape rests on a duripan at 102 cm depth. Clay minerals in the argillic horizon are serpentine, chlorite, and smectite, and predominantly palygorskite. In this soil with more Mg than Ca, there are no carbonate accumulations, which is unusual for an Aridisol.

6.6.8 Ultisols

Ultisols have argillic horizons from which exchangeable bases have been depleted by leaching. They are generally more strongly weathered than Alfisols and are characteristically in warmer climates where weathering is more intense. The Ultisol in Figure 6.1K is in the Udults suborder, because it has a udic soil moisture regime. The natural plant cover of Ultisols is characteristically broadleaf forest, either evergreen or deciduous.

6.6.9 Mollisols

Mollisols are soils with Mollic epipedons, either with or without argillic horizons in the subsoils. The Mollisol in Figure 6.1H is in

Figure 6.5 A fluvial Entisol on a mangrove (*Pelliciera rhizophorae*) tidal flat on the Pacific coast of Costa Rica. A slice from the upper 0.5 m of soil is laid out on the ground.

the Ustols suborder, because it has an ustic soil moisture regime; and it is an Argiustol, because it has an argillic horizon. Mollisols are commonly considered to represent the grassland soils, but they are also common in forests where the soils have xeric or ustic soil moisture regimes; for example, along the Pacific coast in California and Central America.

6.6.10 Alfisols

Alfisols have argillic horizons and generally lack mollic epipedons. The Alfisol in Figure 6.1A is in the Udalfs suborder, because it has a udic soil moisture regime. Alfisols are the characteristic soils of temperate deciduous forests, but they are common in warmer climates also, where the soils are young, or leaching has not been sufficient to remove enough basic cations to lower the base saturation substantially.

6.6.11 Inceptisols

Inceptisols are weakly developed soils that do not have aridic soil moisture regimes and that lack strong expressions of diagnostic features that distinguish the other soil orders. They commonly have cambic horizons and may have umbric or histic epipedons. They can even have mollic epipedons if they lack argillic horizons and the base saturation in a horizon below the mollic epipedon is low (< 50%). Inceptisols are common in recent deposits, such as alluvium and loess, and in mountainous areas where erosion prevents the development of diagnostic features that form slowly. The Entisol in Figure 6.1G is in a mountainous area and lacks diagnostic features other than a cambic horizon.

6.6.12 Entisols

Entisols are rudimentary soils that lack any definitive expression of diagnostic features that differentiate the other soil orders. They occur in all climatic regions, except the polar areas of permafrost. Entisols are most extensive in arid areas that lack plant cover. The largest of these areas is the Sahara Desert (Figure 6.3), where the absence of vegetative cover allows wind to move the surface deposits frequently and sand dunes are common. The Entisol in Figure 6.1D is in a recent alluvial fan deposit of the Mojave Desert, California. It is marginal to a Cambid (Aridisol with a cambic horizon), because the A horizon lacks $CaCO_3$ and it has accumulated in the Ck horizon—only the absence of obvious soil structure in the AC horizons prevents it from being a cambic horizon. In contrast to the dry Entisol of Figure 6.4, the Entisol of Figure 6.5 is wet; it is inundated daily.

6.7 Global Distributions of Soils

The pattern of soil distributions (Figure 6.3) is similar to the pattern of climates (Figure 5.6), except for the Andisols, Histosols, and Inceptisols that have distributions that are not so closely related to climate. The schematic map of Figure 6.3 is based largely on Europe for the northwest part, North America for the northeast part, Africa and South America for the southwest part, and Africa and Southeast Asia for the southeast part, ignoring the rain shadows of the American

Cordillera and high plateaus of Central Asia. The general patterns are obvious, but the boundaries are somewhat subjective. One of the more arbitrary decisions was to include Mollisols in the southern hemisphere to represent the Pampas of South America, which has no counterpart in Africa. The distributions of Andisols and Histosols are discussed in Section 6.6. Inceptisols are rudimentary soils that are found in all of the climatic regions. The largest area of Inceptisols is in the mountains of northeastern Asia.

6.8 Hydric Soils

Soils are a major feature in the identification and delineation of wetlands. Hydric soils are those that formed under aquic conditions of saturation, flooding, or ponding long enough during the growing season to develop anaerobic properties in the upper parts of them. Evidence of aquic conditions includes the reduction of iron, manganese, and sulfur. Features indicative of reduction are called *redoximorphic* features.

Indicators of hydric soils differ from one region to another (Hurt and Vasilas 2006). Deep organic soils or soils with histic epipedons are indicators of aquic conditions. In soils with less organic matter, iron and manganese depletions and concentrations are the common indicators of aquic conditions. Because iron and manganese are colorless in reduced forms, lack of color other than that of the primary minerals in a soil can be an indicator of reduction, although there are other reasons for some soils to lack color. Both ferrous iron (Fe^{2+}) and reduced manganese (Mn^{2+}) are soluble and, if not moved all the way out of a soil, can be concentrated in mottles and nodules. The mottles are produced when concentrated iron becomes oxidized to form yellow or red oxyhydroxides or oxides, or when manganese is oxidized to form black concentrations or nodules. Manganese concentrations are less common than iron concentrations and usually deeper in soils.

Questions

1. Why are phosphorous (P_2O_5) concentrations > 250 Mg/kg of soil associated with human activity?

2. Will the soil temperature and moisture regimes of some soils change with changes in climate? What STRs or SMRs are likely to change on a warming Earth?

3. Are the classifications of some soils with cryic soil temperature regimes likely to change following some land use changes such as clearing boreal forests for permanent pasture?

4. Is a soil with an argillic horizon in a temperate climate more likely to be an Ultisol in a seasonally wet (udic STR) or a seasonally dry (xeric STR) climate?

5. Why are Mollisols more common in grasslands than in forests?

6. Why are Ultisols more common on the eastern sides of continents than on the western sides?

7. How can well-drained, cryoturbated soil in which bleached E horizons are mixed with reddish spodic horizons be differentiated from mottled hydric soils? There are numerous circumstances where the general rules for identifying hydric soils are not applicable.

Supplemental Reading

Buol, S.W., R.J. Southard, R.C. Graham, and P.A. Daniel. 2011. *Soils Genesis and Classification*. Cambridge University Press, Cambridge.

Foth, H.D., and J.W. Schafer. 1980. *Soil Geography and Land Use*. Wiley, New York.

Hurt, G.W., and L.M. Vasilas (eds.). 2006. *Field Indicators of Hydric Soils in the United States: A Guide for Indentifying and Delineating Hydric Soils*. Version 6.0. USDA, Natural Resources Conservation Service, Lincoln, NE.

IUSS Working Group. 2006. *World Reference Base for Soil Resources*. World Soil Resources Report 103. Food and Agricultural Organization of the UN, Rome.

Soil Survey Staff. 1975. *Soil Taxonomy: A Basic System for Making and Interpreting Soil Surveys*. Agriculture Handbook 436. USDA, Washington, DC.

Soil Survey Staff. 1999. *Soil Taxonomy: A Basic System for Making and Interpreting Soil Surveys*. Agriculture Handbook 436. USDA, Washington, DC.

7

Soils and Landscapes

Pedons seldom occur individually; many similar pedons generally occur together in a soil body that is sometimes referred to as a *polypedon*, or simply a soil. They do not occur randomly across a landscape. The kinds and distributions of polypedons (or soils) are controlled by geology, or stratigraphy and lithology, topography, climate and microclimate, living organisms, and historical events. These are commonly referred to as the *factors of soil formation* that determine the kinds of soils that develop through time (Shaw 1930; Jenny 1941; Thorp 1942). Processes of soil formation are discussed by Simonson (1959).

7.1 Lithology and Stratigraphy

Rocks and surficial geologic deposits are the parent materials from which soils develop. The texture, mineralogy, and chemistry of young soils are determined largely by their parent materials. As weathering and leaching proceed, soil properties are less dependent upon parent material, and the influences of soil drainage and climate become more apparent.

Soils that are initially coarse textured because of a high quartz or feldspar concentration in parent materials may remain coarse textured for hundreds or thousands of millennia if quartz predominates, or become fine textured if there is little quartz and weathering is sufficiently intense to destroy the feldspars. At the other extreme, mudstones and shales commonly produce fine-textured soils relatively quickly.

Fertility effects, as with textural effects, are most notable for the extremes. Soils with quartzite, granite, or rhyolite parent materials may lack sufficient basic cations (Ca, Mg, and K) to yield highly productive plant communities, especially in wet climates where leaching removes these cations rapidly. Soils with ultramafic parent materials (peridotite and serpentinite) have little K and so much Mg that the little Ca present is not readily available to plants, and Ni toxicities may be common. Soils with limy parent materials are initially alkaline and remain alkaline or neutral until excess carbonates have been leached from them, which may require moderately wet, or wet, climate. Soils developed from limestone do not have a favorable balance of nutrients for luxuriant plant growth, and in alkaline soils P is fixed in $Ca_3(PO_4)_2$, which is unavailable to plants.

The attitudes of layers, or bedding, in rock strata can be important. For example, soils on slopes that are parallel to a bedding plane are commonly shallow, presumably because water draining through the soils is deflected downslope above bedding planes, causing surface erosion by overland flow of water, or by saturating soils and activating mass failure.

Geologists make many more lithologic distinctions than are important in soils. Lithological distinctions that are generally important in soil and ecological landscape mapping are presented in Appendix D. Although some rocks, such as serpentinite and limestone, are well known for having distinctive soils, it is sometimes difficult to predict before doing field investigations what differences in lithology can be expected to make distinctively different soils. For example, there is a gradual shift in chemistry and mineralogy from granite through quartz-monzonite, granodiorite, quartz-diorite, biotite-diorite, hornblende-diorite, to gabbro, and a parallel sequence from rhyolite to basalt in volcanic rocks. The more mafic rocks, gabbro and basalt, have more basic cations, and the mineral nutrient content is more favorable than in soils derived from the weathering of silicic rocks such as granite and rhyolite. Chemically the choice for separating them in soil mapping is arbitrary; therefore, a break is generally made in the middle of the range for volcanic rocks, separating soils derived from rhyolite and dacite from those derived from andesite and basalt. In the sequence of plutonic rocks, silicic rocks generally have enough biotite that they commonly disintegrate to a coarse sandy mass (like granulated sugar) called grus, whereas mafic rocks lacking biotite do not weather to grus. This has profound effects on soil physical properties and slope stability. Therefore, soils on plutonic rocks are most appropriately split between those on hornblende-diorite and gabbro, on one hand, and those on biotite-diorite and more silicic rocks, on the other. Although soils on volcanic flow rocks may be similar to those on plutonic rocks, they are generally separated from them because of differences in texture, mineralogy, and structure.

Sedimentary and pyroclastic deposits and rocks, such as sandstone and shale, or mudstone, and ash or tuff, can be hard, soft (penetrable with a spade), or unconsolidated. These differences in consolidation or cementation may be sufficient to produce different soils, even where mineralogical compositions are similar.

Soils derived from metamorphic rocks may be similar to those derived from igneous or sedimentary rocks if the metamorphism is slight, or quite different if the metamorphism is great enough to change the igneous or sedimentary progenitor substantially. Preliminary separations of soils may be based on the principles stated here and in Appendix D, but the ultimate decisions as to what parent material differences are sufficient to produce essentially different soils depends upon field investigations. A parent material difference that produces substantially different soils in young landscapes may not be important in older landscapes where soil development converges to a common soil on many different kinds of parent materials.

7.2 Landforms and Topography

Topographic features of all sizes have important influences on the kinds of landforms produced and consequences for their development. Large features such as mountains are notable for atmospheric effects, which are commonly called orographic effects, related to the movement of air across them. More notable on smaller landforms are the effects of topographic controls on the flow of water and dissolved substances through soils and the movement en masse of rocks, regolith, and soils on slopes. Drainage has direct effects on plants by influencing the amount of water that accumulates and the flow of gases in soils. Plants on excessively drained soils may be limited to those plants that are drought tolerant. Plants in poorly drained soils may be limited to those with roots adapted for anaerobic conditions.

Landforms that are constructed by the deposition of glacial drift, sediment from streams or sea currents, or sand from wind, or that are constructed by volcanic activity, have specific names (Table 2.3). Landforms developed by erosion are less distinctive and not identified as specifically. Mountains and hills are commonly erosional landforms, but some are constructed by glacial or volcanic processes. Slopes on erosional mountains and hills are commonly, and

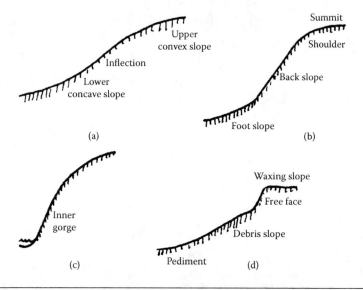

Figure 7.1 Four kinds of slope form in profile (two dimensions). (a) Moderately steep slopes and steep slopes that lack a linear section. (b) Steep and very steep slopes with a linear section between an upper convex and a lower concave section. (c) Inner gorge slopes that are above a stream or floodplain and lack a lower concave section. (d) Slopes with a vertical section of exposed rock and a debris slope below the exposed rock.

somewhat arbitrarily, designated summit, shoulder, side, and foot slopes (Figure 7.1). Sideslopes are sometimes called backslopes where they are much steeper than shoulder and footslopes. Each position on a slope can be characterized by its gradient (degrees or percent), direction (or aspect), shape, and surface form. Shape is described as linear, convex, or concave in downslope and across slope directions; for example, a convex-linear slope is convex in the slope direction and linear across the slope along topographic contours. Nine combinations of slope shape, based on these three parameters in two directions, are possible, as illustrated by Ruhe (1975). Convex-linear, concave-linear, linear-convex, linear-concave, convex-convex, and concave-concave slopes are common, whereas the convex-concave and concave-convex combinations are much less common. Slopes that are linear in both directions are planar (planar slopes are not necessarily level). Rocky, smooth, stepped, hummocky, and pitted are some of the more common surface forms.

Slopes retreat or decline by a combination or various processes. Some of the major processes are rock fall on a free face (Figure 7.1D); landslides in an inner gorge (Figure 7.1C), with very steep slopes

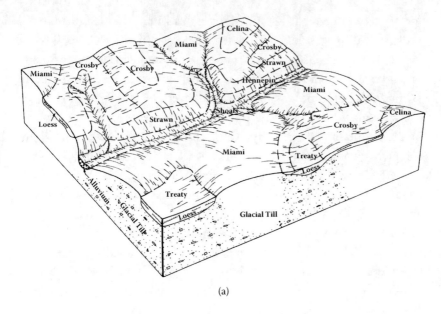

(a)

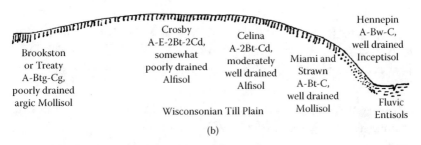

(b)

Figure 7.2 Soils on a late Pleistocene, Wisconsinan, till plain in eastern Indiana. (a) A landscape mapped in a soil survey of Wayne County. (From Blank, J.R., *Soil Survey of Wayne County, Indiana*, USDA, Soil Conservation Service (now Natural Resources Conservation Service), Washington, DC, 1987.) (b) Classification and horizon sequences in the soils.

maintained by a stream cutting downward; erosion largely by overland flow of water, maintaining relatively shallow soils on shoulders and backslopes, especially on shoulders (Figure 7.1B); and erosion by transport of dissolved elements and compounds and colloidal suspensions in water that drains through soils (Figure 7.1A). Combinations of two or more of these processes are important on most slopes.

Bushnell (1943) described topographic sequences of soils based on drainage patterns across the late Pleistocene glacial till plains of Indiana. A typical sequence on Wisconsinan till is illustrated by a landscape in eastern Indiana (Figure 7.2) that was cleared of broadleaf forest in the nineteenth century for agriculture. The parent material is dense, calcareous, loamy till with a thin cap of loess. Enough water has moved downward to leach free carbonates to depths below 50 cm, or to the C horizons, and some translocated clay has accumulated in B horizons, but some water has drained overland and laterally through the soils, above the compacted till. Slopes gradients across the summit of the till plain are generally < 2%, and the soils (Crosby series) are somewhat poorly drained Alfisols with A-E-2Bt-2C horizons, with moderately well-drained soils (Celina series) near the undissected edges of the till plain surfaces. (The number 2 antecedent to the B and C horizon designations indicates that there are two soil parent materials, which are loess over glacial till in this case.) Dissection has produced a valley with steep sideslopes occupied by Inceptisols (Hennepin series) and a floodplain with fluvic Entisols. Well-drained Alfisols (Miami and Strawn series) that have lost their loess caps from erosion occupy gentle to moderately steep slopes above the valleys. Loess and till that have washed into depressions in the till plain are occupied by poorly drained, argic Mollisols (Brookston and Treaty series). Organic matter has accumulated in some of the depressions to produce Histosols (Houghton series). Relief from the Crosby to the Brookston soils is commonly < 1 or 2 m.

Soils on the late Pleistocene, Wisconsinan, till plain are not old enough to show great differences in clay mineralogy. A much older landscape in the summer-dry coastal hills of southern California has a sequence of soils that do differ greatly in clay mineralogy (Figure 7.3).

The sequence is in granitic terrain and includes Vista soils on summit and shoulder slopes, Fallbrook soils on sideslopes, Bonsall soils on footslopes, and Bosanko soils in alluvium at the bottom. Loss of material by erosion from summit and shoulder slopes has prevented the development of deep soils with argillic horizons there; Vista soils there have A-Bw-Cr profiles. Biotite (mica) in the parent material has been hydrated to produce vermiculite (V) in the Vista soils. Below the shoulders, material has been carried across sides-lopes with less soil loss from sideslopes than from shoulder slopes, allowing the development of argillic horizons; Fallbrook soils there have A-Bt-Cr profiles. Leaching of cations from the Fallbrook soils has created an environment in which feldspars from the parent material weather to produce kaolinite (K). Sediments washed across

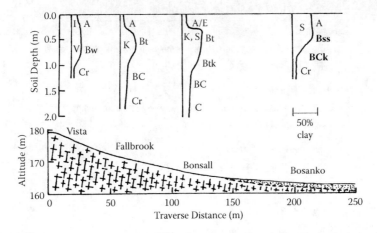

Figure 7.3 A transect across soils in granitic terrain, San Diego County, California. The predominant clay minerals are illite, or hydrous mica (I), vermiculite (V), and kaolinite (K) in the intensively leached soils in the upper part of the landscape and smectite (S) downslope where silica and basic cations have accumulated in the soils. (From Nettleton, W.D. et al., *A Toposequence of Soils in Tonalite Grus in the Southern California Peninsular Range*, Soil Survey Investigations Report 21, USDA, Soil Conservation Service, Washington, DC, 1968.)

the sideslopes have accumulated on footslopes and extra water from upslope has leached the footslope surface soils to produce eluvial horizons, but slow drainage through the soils has led to the accumulation of carbonates in the subsoils; Bonsall soils on the footslopes have A-E-Bt-Btk-BC-C profiles. In alluvium at the bottom of the sequence, the accumulation of silica and salts from drainage water has promoted the formation of smectites (S), which shrink and swell to churn the soils and prevent much horizon differentiation; Bosanko soils have A-Bss-BCk-C profiles.

Another example of profound soil and plant community differences in a landscape with little topographic relief is described in Section 7.7. Also, Swanson et al. (1988) have some examples of landform effects.

7.3 Climate

The main elements of climate are solar insolation, air, and water. These factors have been introduced in Chapters 4 and 5. The common measures of climate are precipitation and air temperature. Precipitation is the amount of water in both rainfall and snowfall. Wind is a major element of weather, and wind patterns are important determinants of global climates.

Table 7.1   Zonal Soil Development in Different Climates or Biomes

Polar
Dry Entisols → inorganic Gelisols (Orthels; for example, Anhyorthels)
Moist Entisols → organic Gelisols (Histels)

Boreal Forest
Dry, mixed forest, Entisols → Inceptisols → Alfisols (uncommon)
Moist, conifer forest, Entisols → Spodosols

Temperate Forest (broadleaf)
Entisols → Inceptisols → Alfisols → Ultisols

Temperate Grassland
Entisols → Mollisols

Tropic of Cancer and Tropic of Capricorn Deserts
Entisols → Aridisols

Deciduous Tropical Forest
Entisols → Mollisols

Semideciduous Tropical Forest
Entisols → Inceptisols → Alfisols

Evergreen Tropical Forest
Entisols → Inceptisols → Ultisols → Oxisols

Perudic (perpetually wet or moist) Tropical Forest
Entisols → Inceptisols → Oxisols

Entisols are azonal soils, occurring in all climatic zones. Intrazonal
soils differ from zonal soils, because of parent material or topographic influences (relief and drainage): they are most Andisols and
Vertisols, many Histosols, and some subgroups in other soil orders.

Water is important for hydration and other chemical weathering
reactions, activity of organisms in soils and their plant communities,
and transformation and translocation of materials. Temperature indicates the intensity of heat (Appendix A). It controls the kinds and
rates of many chemical and physical reactions and the activities of
organisms in soils.

Distributions of some soils are largely dependent upon climate.
They occur in zones that are commonly related to distance from the
equator or from an ocean (Table 7.1, Figure 6.3). Very cold to extreme
climates near and above the Arctic Circle (66.5°N) are the provenance
of Gelisols, including frozen organic soils. They are common in northern Asia and northern North America. The Antarctic continent is dry

and lacks the vegetation to develop organic soils, although mosses and lichens grow sparingly within a few degrees of the South Pole. Boreal climates, between about 50 and 60°N, are conducive to development of Spodosols, and Histosols are common. Spodosols are sparse in the southern hemisphere, because there is little land at latitudes > 45°S. Alfisols develop at mid-latitudes and adjacent to the zones of Aridisols that are centred on the Tropics of Cancer (23.5°N) and Capricorn (23.5°S). Alfisols in Central Europe (and eastern North America) are Udalfs (moist summers), those in Mediterranean climatic areas are Xeralfs (dry summers), and those in tropical areas are mostly Ustalfs (plant growing seasons with both moist and dry periods). Mollisols are common at mid-latitudes, also, but where there is less precipitation inland from the wetter Alfisols. There is a broad east-west zone of Mollisols on the steppes of Central Asia that extends eastward into southeastern Europe. Oxisols occur in the tropical zone. Ultisols are commonly in zones between Alfisols and Oxisols, where they are less intensely weathered than Oxisols and more intensively leached of basic cations than Alfisols, but parent materials and ages of landforms are also very important in the distributions of Ultisols. Distributions of Vertisols are dependent largely on parent materials and relief, or drainage, but they do require a seasonally dry climate. These conditions for the development of Vertisols are common in the Upper Nile River Basin of Africa, in Central India, in Eastern Australia, in the lower Parana and La Plata River Basins of Argentina and Uruguay, and in Texas and Mexico on the coastal plain of the Gulf of Mexico.

Zoning of soils by latitude that is readily discernible from Eastern Asia and Europe down through Central Africa is altered in North America by high mountains along the west side of the continent and by relatively recent (very late Pleistocene to Holocene) till, outwash, and aeolian (loess) deposits from areas of extensive continental glaciation on the northeast. Gelisols are mostly near or above the Arctic Circle, as expected, and Spodosols dominate the boreal zone between about 45 or 50 and 60°N. Histosols are common along with the Spodosols. The large zone of Mollisols on the Great Plains of North America is oriented north-south in the rain shadow of the western Cordillera. The more moist Mollisols (Udolls) are east of the drier Mollisols (Ustolls) in North America, whereas the Udolls are north of the Ustolls in Asia. Another unique facet of soil distributions in eastern North America is that the boundary between Alfisols and Ultisols is practically coincident with the limit of the last continental glaciation. The approximately 15 to 100 ka since the retreat of Wisconsin glaciers has not been long enough for the soils of the glaciated terrains to have been sufficiently depleted of basic cations to be Ultisols. The implication is that given more time, Ultisols will develop farther north, above 40°N, in humid climates of mid-latitudes. They actually do occur above 42°N in the Willamette Valley, and even north of the Columbia River toward Puget Sound.

Trade winds blowing across cold currents along the western coasts of the continents are largely responsible for the arid climates centred about the Tropics of Capricorn and Cancer. Because the winds blow across cold currents, they take up little moisture from the oceans. These dry winds are particularly prominent in the southern hemisphere where the Humboldt Current along the coast of South America and the Benguela Current along the coast of southern Africa are fed from the Antarctic Circumpolar Current. The Kuroshio Current along the eastern coast of Asia and the Gulf Stream along the eastern coast of North America are warmer, supplying moisture to the continents and making the climates too wet for Tropic of Cancer deserts. Tropics of Capricorn and Cancer deserts are large in Australia, and very large from northern Africa into western Asia, where there is more land mass. Tropic of Cancer deserts in North America are modist in comparison to other larger and drier Tropics of Cancer and Capricorn deserts.

Large mountains that force the wind to blow over them cause regional climates to differ from the usual global patterns. The Sierra Nevada in California is an example of a mountain range that causes these orographic effects. A transect across the central Sierra Nevada at about 37°N latitude where the mountains reach heights about 4000 m is shown in Figure 7.4, with data at sites along the transect tabulated in Table 7.2. Annual precipitation is < 300 mm in the centre of the Great Valley, on the west side of the Sierra Nevada. As the prevailing westerly winds rise to cross the mountains, the air cools and loses moisture as precipitation. Temperatures decline and precipitation increases with elevation, up to about 3000 m altitude. As the air that has been depleted of moisture in rising to cross the Sierra Nevada descends east of the Sierra Nevada, it becomes warmer and drier. From west to east, the relatively mature soils are (1) warm Alfisols in grasslands on glacial

Figure 7.4 A transect from the Great Valley of California across the Sierra Nevada. The transect site characteristics are listed in Table 7.2.

outwash and alluvial fans from the Sierra Nevada; (2) warm Alfisols in oak woodlands on the foothills; (3) cool Alfisols in mixed conifer forests; (4) cool Inceptisols with umbric epipedons in conifer forests; (5) cold Inceptisols with umbric epipedons in fir forests; (6) very cold, commonly shallow, Inceptisols with umbric epipedons in lodgepole pine forests; (7) very cold, generally shallow, Entisols with herbaceous vegetation on mountain summits; (8) very cold, commonly shallow, Entisols in whitebark pine plant communities; cold Mollisols lacking argillic horizons in Jeffrey pine/bitterbrush plant communities (not represented in Table 7.2); and (9) cold Mollisols with argillic horizons in big sagebrush–bitterbrush plant communities on some of the older (>100 ka) glacial moraines. Klyver (1931) has described the vegetation patterns across the Sierra Nevada at about 37°N latitude in more detail, Barshad (1966) has identified clay minerals in many of the soils, and Dahlgren et al. (1997) have more details on the soils in a transect up the western slope of the central Sierra Nevada. Drier sites in the rain shadow east of the Sierra Nevada are represented in Table 7.2 by soils and plant communities on alluvial fans from the White Mountains:

Table 7.2   Landscape Characteristics of Twelve Sites along a Transect from the Great Valley Eastward across the Central Sierra Nevada to the White Mountains, California

Site No.	Alt. (m)	Map mm	Geologic Material	Slope/ Landform	Soil Class/ Horizons	Plant Community
1	130	300	Granitic alluvium	Alluvial fan	Warm (duric) Alfisol A-Bt-(Bq)-Cr	Wild oat–soft brome
2a	670	500	Granitic rock	Moderately steep hills	Warm Alfisol A-Bt-Cr	Blue oak/wild oat–soft brome
2b	920	750				Gray pine/interior live oak
3	1180	800	Granitic rock	Moderately steep hills	Cool Alfisol Oi-A-Bt-Cr	Ponderosa–incense cedar–white fir/black oak manzanita
4	1950	900	Granitic rock	Steep mountains	Cool umbric Inceptisol Oi/ Oe-A-Bw-Cr	Ponderosa–incense cedar–sugar pine–white fir/black oak/com. Manzanita–creeping snowberry
5	2300	1000	Granitic rock	Steep mountains	Cold umbric Inceptisol Oi/ Oe-A-Bw-Cr	Red fir–white fir/bush chinquapin–greenleaf manzanita–pinemat manzanita–mountain whitethorn
6	2820	1100	Granitic rock	Moderately steep ridge	Very cold lithic umbric Inceptisol Oi-A-Bw-R	Lodgepole pine/dwarf lupine/needlegrass
7	3600	1000	Metamorphic rock	Sloping ridge	Very cold lithic Entisol A-C-R	Cushion phlox–cushion draba
8	3190	800	Granitic rock	Moderately steep ridge	Very cold lithic Entisol Oi-A-C-R	Whitebark pine/bush chinquapin–mountain pride–white heather
9	2180	400	Mixed till	Moderately steep hills	Cold argic Mollisol Oi-A-Bt-C	Big sagebrush– bitterbrush–roundleaf snowberry/needlegrass
10	2240	300	Mixed alluvium	Alluvial fan	Cold argic Aridisol A-Bt-Ck	Pinon–juniper/big sagebrush–bitterbrush
11	1860	200	Mixed alluvium	Alluvial fan	Cool calcic Aridisol A-Bwk-Ck	Spiny mendora– shadscale–Nevada joint fir
12	1195	120	Mixed alluvium	Alluvial fan	Warm sodic argic Aridisol A-Btn-C	Black greasewood–big Saltbush–bud sagebrush

Abbreviations: Alt., altitude; MAP, mean annual precipitation. Plant taxonomic names are in Appendix F.

(10) cold Aridisols with argillic horizons in piñon–juniper woodlands; (11) cool Aridisols with cambic horizons containing accumulations of Ca-carbonates in shadscale desert; and (12) warm Aridisols with argillic horizons containing accumulations of soluble salts and supporting black greasewood and other desert shrubs, and with clumps of Indian ricegrass (*Achnatherum hymenoides*) on low mounds of aeolian sand and silt that accumulated around shrubs.

In the Sierra Nevada, the elevation trend is to greater soil leaching and organic matter accumulation with greater elevation and precipitation, and slower weathering and clay mineral formation. The first clay mineral to form from weathering of granitic rocks, which are the dominant rocks in the central Sierra Nevada, is commonly vermiculite (Ismail 1970). Soils at the higher elevations lack the clay minerals to form argillic horizons. Climatically related clay minerals are allophane and imogolite above 1500 m, halloysite about 1000 to 1500 m, and smectite below 1000 m. Recent analyses indicate the dominance of kaolinite in very old soils of the Great Valley, which is suggestive of wetter past climates with greater leaching of basic cations. Diminished precipitation in the rain shadow east of the Sierra Nevada has not leached all Ca-carbonates from the subsoils, and many of the soils (not represented in Table 7.2) have accumulations of silica; commonly the soils have calcic horizons or duripans.

Locally, slope aspect and wind direction can have profound effects on the soils and vegetation. In semiarid areas of the mid-latitudes in western North America, the prevailing westerly winds sweep snow over mountain and hill summits and drop it on leeward slopes, where the extra moisture from melting snow leaches the soils more thoroughly on east-facing slopes and allows aspen trees to grow where there are none of west-facing slopes (Glassey 1934).

7.4 Biota (Plants, Animals, and Other Living Organisms)

Soils are open systems. Initial soil systems are colonized by organisms from surrounding areas, and the species compositions change as species migrate to and from soil systems. Thus, the soil biota is a temporally variable factor of soil formation.

The greatest impact of the soil biota is in the production of organic matter and in mediating the transformations of it (Chapters 8 to 10).

Plant roots and animals have great effects on soils belowground. The chemical environments and microbial populations around roots, which is referred to as the rhizosphere, differ greatly from those of soil beyond the rhizosphere. Both vertebrate and invertebrate animals move large volumes of soils (Butler 1995; Goudie 1988; Hole 1981; Johnson 1990; Lobry de Bruyn and Conacher 1990). The work of animals that bring soil materials to the surface and build mounds is most easily observed and quantified, although many more animals live below and their presence is not evident at the surface. The magnitudes of earth moving are greatest for earthworms, and diminish through ants and termites to vertebrates (reptiles, birds, and mammals). Earthworms add an average of 1 to 10 kg/m² of soil to mounds each year in many areas, and up to about 25 kg in some areas.

7.5 Plant Communities and Ecosystems

Plant community distributions depend upon all of the soil-forming factors, and upon the existing soils. Each combination of soil and plant community is a different ecosystem, or geoecosystem. Although some definitions of *ecosystem* include soils and their parent materials, the term *geoecosystem* ensures that soils and their parent materials are included in the spatial (three-dimensional) systems. Although terrestrial geo-ecosystems can generally be defined by their soil parent materials, landforms, soils, and plant communities, each geoecosystem also has characteristic suites of microorganisms, fungi, and animals.

Plant communities can be recognized by the general character, or physiognomy, of the vegetation, by the dominant species, or by a combination of these features. Physiognomy is particularly useful at regional scales, where grasslands are generally in drier areas and forests are generally in wetter areas. Within regions, however, different soils may have distinctly different kinds of plant communities; for example, grasslands may occur on limestone soils surrounded by forests on adjacent soils with other parent materials. It is very important to consider rock types and soils in predicting local distributions of plant communities.

7.6 Soils and Time

Time is just a concept, rather than a thing that has substance. It is an interval between events—either between two events in the past or between an event in the past and the present. Time is measured by the duration of a cyclic event, such as the rotation of the earth about its axis or the swing of a pendulum, or by the distance travelled by an object moving with constant velocity. A second is defined as 9,192,631,770 oscillations of electrons in a cesium atom, although physicists are not content with this accuracy and are seeking ways to measure fractions of seconds (nanoseconds) more precisely.

Both cyclic and noncyclic changes occur in soils. Soil temperature is an example of a property that changes in both diurnal and annual cycles (Chapter 4). The same, or similar, temperatures recur daily and each year. Some soil properties change both cyclically and progressively; for example, soil nitrate concentrations fluctuate annually but may increase substantially over the years or decades of initial soil development following the exposure of till by a retreating glacier and colonization of the till by plants and other living organisms.

Some of the progressive (and retrogressive) changes in soils are permanent translocations by leaching, accumulations of clays and other substances, and alteration of minerals. Directions and rates of these changes are highly dependent on local environments, particularly on the nature and magnitudes of the soil-forming factors. For example, Ca carbonates may accumulate in some soils and be leached from others where the precipitation is greater. In soils where carbonates do accumulate, it may take a thousand years for enough to accumulate to make a calcic horizon, or it may even take many thousands of years in drier soils or in soils with lower Ca concentrations. It is very difficult to predict the rates at which diagnostic horizons will form with any more precision than an order of magnitude (a factor of 10).

Rudimentary soils, lacking horizon development, are Entisols. As plant detritus collects on soils and some of the organic matter is incorporated in to the soils, A horizons develop quickly, and in a few hundred years or less, they may become mollic or umbric epipedons. Thus, in a few hundred years Entisols may become Mollisols in dry climates, Inceptisols with umbric epipedons in more highly leached soils, or Andisols with melanic epipedons in soils with amorphous materials. Spodosols may form in a few hundred years also, but only in wet climates with specific kinds of vegetation, such as conifer forest, for example.

In areas where the climate, topography, and vegetation are not conducive to the development of mollic or umbric epipedons, Inceptisols with cambic horizons may develop in a few thousands years, or calcic horizons may develop in drier areas. It generally takes many thousands of years for argillic horizons to develop, commonly tens of thousands of years. In eastern North America where land was exposed only after the last continental glacier retreated about 10 to 15 thousand years ago, Alfisols have formed with argillic horizon. Soils nearby that have not been glaciated are more weathered and leached, generally with more kaolinite clays and redder colors, and have formed Ultisols. In western North America, where ranges of precipitation are greater, rates of leaching are more important than soil age in differentiating between Alfisols in drier climates and Ultisols in wetter climates. Very old soils, hundreds of thousands to millions of years old, have highly weathered horizons, or oxic horizons, that make them Oxisols. Oxisols are most common in hot, wet climates, where they form more rapidly than in drier or cooler climates.

Vertisols form in hundreds of years, or less, in parent materials that were initially high in swelling clays, or in somewhat longer time in basins where the acquisition of swelling clays is dependent upon the accumulation of the basic cations and silica that promote the development of smectite clays from the weathering process. They are sparse or absent in well-drained soils where there has been enough precipitation to leach most of the basic cations from the soils.

Organic soils, or Histosols, occur where production rates of organic detritus are greater than rates of decay. Decay of organic detritus is retarded in wetland soils, because oxygen diffuses very slowly through soils that are saturated permanently, and anaerobic decay is relatively slow. Cool temperatures retard decay of organic detritus also. Therefore, accumulation of organic detritus is more rapid in cool climates than in warm ones. Organic soils are generally common in areas where there are Spodosols. In cool, wet climates, organic detritus typically accumulates at rates up to about 1 mm/year. Therefore, the formation of a Histosol 40 cm thick requires at least 400 years. Because rates of decay are relatively rapid in drained (unsaturated) soils, only in cool and damp, yet productive, environments, such as southeastern Alaska, do organic soils form that are not continuously saturated with water. The organic soils, or Histosols, in well-drained soils are called Folists, because the organic detritus is predominantly leaves that are in various stages of decomposition.

7.7 Soil and Ecological Landscape Maps

Ecological land management requires spatial data. It is not enough to know the geologic, pedologic, and biotic characteristics of individual sites and pedons. Data from many sites and pedons must be grouped into manageable numbers of landscape units that respond similarly to management. The distributions of the landscape units must be known in order to plan for effective and efficient land management. Management requires knowledge of the capabilities of each parcel of land and its limitation to management. This knowledge is conveyed by soil maps and by ecological landscape maps and reports that indicate the capabilities and limitations of map units.

Although a pedon is small, many contiguous pedons that are similar to one another comprise a soil that can be considered a unit of ecological land management. The boundaries between soils having pedons in different soil classes are generally geologic, landform, or natural plant community boundaries.

One can study soil physics and chemistry without much field experience, but to learn how soils are related to other landscape features, there is no better way than to map soils and other landscape components together as integrated systems. Mapping of landscape units that are based explicitly on combinations of lithologic, topographic, soil, and vegetation features has been practiced in a few cases. Pedologists and other land classifiers developed an integrated approach to land classification in Michigan in about 1930 (Schoenmann 1923; Veatch 1937), but integrated approaches to soil survey and landscape mapping remained exceptions, rather than common practices. About the middle of the twentieth century, methods were developed to map land units and systems in Australia based largely on topography, soils, and vegetation (Christian 1957), and Canadians developed guidelines for "biophysical" land classification (Jurdant et al. 1975). The Australian methods were adapted for mapping in the western United States (Wertz and Arnold 1972) and applied to soil surveys in some regions of the U.S. Forest Service. Two examples of integrated ecological landscape surveys are those that were done on the Targhee and on the Bridger-Teton National Forests (Bowerman et al. 1997; Svalberg et al. 1997). A more detailed integrated ecological landscape survey was done on the Blacks Mountain Experimental Forest in California

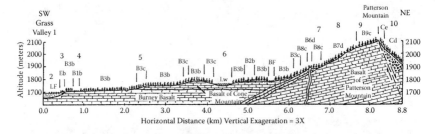

Figure 7.5   A transect across landscape map units of the Blacks Mountain Experimental Forest in the Modoc sector of the Trans-Cascade Volcanic Plateau in the Basin and Range Province of western North America. The map unit characteristics are listed in Table 7.3.

(Alexander 1994). A transect across the ecological landscape map units of the Experimental Forest is shown in Figure 7.5, with descriptions of the featured map units in Table 7.3.

An example of ecological landscape mapping on the basalt plateau of Blacks Mountain Experimental Forest (Figure 7.6) shows how major soil and plant community differences can be related to small differences in topographic relief (map unit descriptions are in Table 7.3). The transition from a wet meadow (map unit Lw) at about 1772 m asl to a highly productive forest (units B3b and B3c) is sharp, being less than 1 m in elevation and only a few meters laterally. Openings in the forests of units B3b and B3c are from logging. The shallow soil of

Table 7.3 A Transect across Landscape Map Units in the Blacks Mountain Experimental Forest on the Modoc Sector of the Trans-Cascade Plateau

Site No.	Alt. (m)	Map (mm)	Soil PM	Slope/ Landform	Soil Class/ Horizons	Plant Community
1 L2	1698	400	Lacustrine sediments	Level, flat, dry lake plain	Poorly drained Vertisols A-ACss-C	Spikerush–mat muhly
2 LF	1699	400	Lacustrine sediments	Level, flat, dry lake plain	Imperfectly drained Alfisols A-Bt-C	Silver sagebrush/ smooth brome–bluegrass
3 Eb	1702	450	Aeolian sediments	Gently sloping inactive dunes	Haplic Alfisols A-Bt-C	Big sagebrush/ velvet lupine/ Idaho fescue
4 B1b	1760	500	Basalt	Slight depression on plateau	Lithic Alfisols A-Bt-Cr-R	Juniper/low sagebrush/ bluegrass
— B2b	1782	500	Basalt	Undulating plateau	Moderately deep stony argic Mollisols	Yellow pine/big sagebrush– bitterbrush
5 B3b	1790	500	Basalt	Undulating plateau	Deep argic Mollisols Oi-A-Bt	Yellow pine/ bitterbrush/ Idaho fescue
— B3c	1785	500	Basalt	Slope on plateau	Moderately deep argic Mollisols	Yellow pine/ bitterbrush/ mahogany
— L2	1777	500	Sediment/ basalt	Level, flat basin fill	Poorly drained Vertisols	Spikerush–mat muhly
6 Lw	1770	500	Lacustrine sediments	Level, flat lake plain	Very poorly drained Vertisols A-Bss-Cg	Rush–tufted hairgrass–false downingia
7 B8c	1940	600	Basalt	Sloping mountainside	Andic argic Mollisol Oi/Oe-A-Bt-C	Yellow pine–white fir/creeping snowberry
8 B7d	2000	650	Basalt	Moderately steep mountainside	Stony argic Mollisol Oi-A-Bw-R	White fir–yellow pine/creeping snowberry
9 B9c	2090	700	Basalt	Sloping summit	Argic Mollisols Oi/Oe-A-Bt-R	White fir/creeping snowberry
10 Ce	2030	700	Basalt colluvium	Steep fault-line scarp	Argic Mollisol Oi/Oe-A-Bt	White fir/creeping snowberry

Abbreviations: Alt., altitude; MAP, mean annual precipitation (approximate). All soils are cold (frigid STR). Yellow pine stands generally contain both ponderosa and Jeffrey pine. Plant taxonomic names are listed in Appendix F.

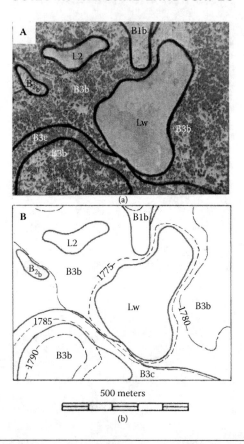

(a)

500 meters

(b)

Figure 7.6   A landscape on the Blacks Mountain Experimental Forest, near the middle of the Figure 7.5 transect. (a) An aerial photograph showing contrasting soil units and plant communities. (b) A topographic map of the same area with elevation contours labeled in meters asl.

unit B1b in a broad trough is more readily delineated from contrasting vegetation than from the minor topographic relief. The lacustrine fill in unit L2 is not as deep as that in unit Lw, allowing more drainage through the underlying basalt and a shorter period of ephemeral wetness in unit L2 that is reflected in the vegetation that is quite different from that in unit Lw.

Questions

1. Are landslides more prevalent where ground slopes are parallel to bedrock bedding planes or where slopes cut across bedding planes?

2. Why do coarse-grained biotite granites weather to granular grus, but not fine-grained rhyolites with practically the same chemical composition and also with biotite?

3. Which of the following landforms are associated with glaciation, which are associated with streams and rivers, and which are aeolian features?

> Moraine
> Floodplain
> Terrace
> Sand dune
> Outwash plain

4. Which of the following deposits are associated with glaciation, which are associated with streams and rivers, and which are aeolian features?

> Loess
> Till
> Alluvium

5. Why is the greatest orographic precipitation in the Sierra Nevada of California at about 3000 m altitude, rather than near the 4000 m summit?

6. Amorphous clay (allophane) and those clays with poorly ordered crystal structures (imogolite) are more common in cool areas, and clays with more orderly crystal structures, such as smectite and kaolinite, are more common in warmer areas. Why?

7. Why are patterns of soil distribution more complex in mountains than on plains? Many soil maps of large areas (small-scale maps) have labels such as simply "mountain soils" for the polygons of mountainous areas.

8. Why is the biota factor of soil formation so controversial? Hans Jenny contended that it consisted of the biogenic forms that were present at "time zero" of soil formation, but soil systems are open and organisms migrate from area to area.

Supplemental Reading

Arnold, R.W. 2006. Soil survey and soil classification. In S. Grunwald (ed.), *Environmental Soil-Landscape Modeling*, pp. 37–59. CRC/Taylor & Francis, Boca Raton, FL.

Birkeland, P.W. 1999. *Soils and Geomorphology*. Oxford University Press, New York.

Brown, D.J. 2006. A historical perspective in soil-landscape modeling. In S. Grunwald (ed.), *Environmental Soil-Landscape Modeling*, pp. 61–103. CRC/Taylor & Francis, Boca Raton, FL.

Jenny, H. 1941. *Factors of Soil Formation*. McGraw Hill, New York.

Ruhe, R.V. 1969. *Quaternary Landscapes in Iowa*. Iowa State University Press, Ames.

Ruhe, R.V. 1975. *Geomorphology*. Houghton Mifflin, Boston.

Schaetzl, R.J., and S. Anderson. 2005. *Soil Genesis and Geomorphology*. Cambridge University Press, New York.

van Breeman, N., and P. Buurman. 2002. *Soil Formation*. Kluwer, Dordrecht, The Netherlands.

8

Primary Production and Plant Nutrition

All living organisms require a source of energy to run their engines, fuelling the metabolic processes that keep them active and promoting their growth. The primary sources of energy are solar radiation and inorganic chemical reactions such as the oxidation of hydrogen sulfide, ammonium, and ferrous iron. The main source is radiant energy from the sun—only specialized bacteria can tap inorganic chemical reactions as the primary source of energy.

Photosynthetic bacteria and algae in the oceans and plants on land are the primary producers of organic products that can be used by animals. Animals are dependent on the primary producers to harness solar energy in the production of food for them. No animals can produce their own food; they consume plant products or other animals.

Most of the terrestrial plant production is by vascular plants with roots in soils. Mosses and other bryophytes are more productive than vascular plants only on very cold, wet soils where plant productivity is much less than on warmer soils. The productivity of vascular plants is highly dependent on the soils from which they obtain water and nutrients, and the vitality of biotic communities depends on plant productivity. This chapter is devoted to the realm of plants.

8.1 Primary Productivity

Green plants have the ability to use energy from the sun to combine water from the soil and carbon dioxide from the atmosphere to make carbohydrates (Figure 8.1).

$$6CO_2 + 6H_2O + energy \rightarrow C_6H_{12}O_6 + 6O_2$$

This process of C fixation is called *photosynthesis*. Excess O_2 from photosynthesis is used in other plant metabolic processes or returned to the atmosphere.

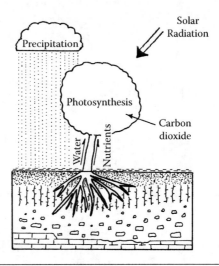

Figure 8.1   Carbohydrate production in green leaves by photosynthesis. Leaves fall to the ground and accumulate in a litter layer over the soil.

The carbohydrate production by photosynthesis is gross primary productivity. It is called primary because the carbohydrates are produced from inorganic constituents. Plants consume carbohydrates in respiration and other metabolic processes. The difference between the amount of carbohydrates produced, which is the gross primary productivity, and the amount used in the metabolism of the plants is the net primary productivity (NPP). It is dependent upon many factors—physical environmental factors, symbionts, and antagonists such as herbivores. Ignoring antagonists, NPP is dependent on the kinds and conditions of plants in an ecosystem or landscape and the portion of the area that they occupy, the supply of water, the amount of carbon dioxide in the atmosphere, the supply of nutrient elements from soils, solar radiation, air and soil temperature, atmospheric circulation, and in many cases, microorganisms that infect roots and aid in the uptake of water and plant nutrient elements from soils.

8.2 Plants in the Physical Environment and Photosynthesis

The effectiveness of photosynthesis in converting solar radiation to carbohydrates is dependent upon air temperatures and the availabilities of water and nutrient elements. Less than 1% of the solar radiation reaching Earth is converted to organic stores of energy, with maxima

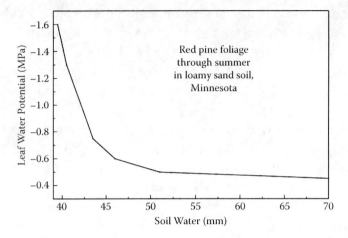

Figure 8.2 Predawn water pressure, or stress, in red fir foliage in relation to soil water content.

of 6% over short periods (Etherington 1982). Rates of photosynthesis increase as air temperatures increase above 0°C. At temperatures of about 30°C, or somewhat higher for tropical plants, plant respiration offsets photosynthesis, and the rate of primary production declines at higher temperatures.

When temperatures are favorable, available water is a major factor limiting photosynthesis. In well-drained soils, photosynthesis is maximum at or near field capacity and declines rapidly as the water potential approaches −1.5 MPa, but the rate of respiration declines less rapidly (Etherington 1982). The amounts of soil water extracted by plants at high leaf water stresses, or potentials more negative than about −0.6 to −1.0 MPa, are minimal (for example, Figure 8.2).

Plants that thrive in wet soils lacking free oxygen are those that are able to move oxygen most effectively to their roots for root respiration (Etherington 1982). A secondary function of O_2 in roots is the oxidation of mobile and toxic divalent forms of Fe and Mn to insoluble and immobile Fe^{3+} and Mn^{4+} in and around the roots.

Photosynthesis in plants consists of (1) light reactions based on the absorption of blue light by chlorophyll a and red light by chlorophyll b, allowing green and yellow light to pass through plants or be reflected from them, and (2) dark reactions that do not require light. Light reactions occur in chloroplasts that are located in the mesophyll of leaves. In the light reactions, water is split to supply electrons for the reduction of $NADP^+$ to NADPH (nicotinamide adenine dinucleotide phosphate) and ATP (adenosine triphosphate) is synthesized. Oxygen (O_2) is a by-product of the light reactions. There are two major paths of photosynthesis, depending on whether 3- or 4-carbon compounds are produced in the light reactions. They are designated the C3 and C4 pathways. In the dark reactions, CO_2 derived from the 3- or 4-carbon compounds is used to produce sugar (glucose) in a carbon reduction cycle that is commonly called the Calvin cycle. In the C3 pathway, both light and dark reactions occur in leaf mesophyll. In the C4 pathway, light reactions occur in leaf mesophyll and dark reactions in bundle-sheath cells surrounding leaf veins. An enzyme used in the Calvin cycle is susceptible to oxidation in a process called photorespiration that reduces the efficiency of photosynthesis. One advantage of the C4 pathway is that employing the Calvin cycle in sheath-bundle cells, rather than in leaf mesophyll, minimizes photorespiration. A third photosynthetic pathway has evolved in succulent plants of arid climates. It is called the pathway of crassulacean acid metabolism (CAM), after the Crassulaceae, which is a common plant family with many desert plants. Its advantage is that the stomata open at night and CO_2 is stored as an acid from which CO_2 can be recovered for daytime photosynthesis; then the dark reactions can occur at night.

The C3 pathway is much more common than the C4 or CAM. It is most efficient in cool, moist climates. The C4 pathway is common in plants of hot, dry climates, especially in grasses, but in plants of many other families also. The optimum temperature for photosynthesis is higher in the C4 than in the C3 pathway. An advantage of C4 pathway is that CO_2 is fixed more rapidly in its light reactions than in those of the C3 pathway, and CO_2 is conserved, requiring less opening of the stomata, and thus conserving water. Besides greater water use efficiency, an advantage of the C4 pathway is that it can operate at higher temperatures than the C3 pathway and produce more biomass. The potential productivity of C4 plants is greater than that of C3 plants. C3 plants can convert only 4.6% of the solar radiation to carbohydrates, compared to 6% for C4 plants (Zhu et al. 2008).

The carbon isotope signatures of organic compounds produced by the C3 and C4 pathways are different. The stable isotopes of carbon are ^{12}C and ^{13}C, and they occur in the atmosphere in ratios about 99 to 1. Plant use favors ^{12}C, so the $^{13}C{:}^{12}C$ ratio is lower in plant biomass than in the atmosphere. The lowering of the $^{13}C{:}^{12}C$ ratio relative to that of a standard, which is benthic foraminifera, is labeled the $\delta^{13}C$, and its units are parts per thousand (‰). Plants utilizing the C3 pathway produce organic compounds with $\delta^{13}C$ values of about $-27‰$ (-22 to $-40‰$ range), and those utilizing the C4 pathway produce organic compounds with $\delta^{13}C$ values of about $-12‰$ (-9 to $-19‰$ range) (Jones 1983). The carbon isotope ratios of plant biomass that is incorporated into soils become the carbon isotope ratios of soil organic carbon. These ratios are retained in the organic matter of buried soils and are indicators of climatic conditions in the past, before the soils were buried. Lower ratios are indicative of warmer temperatures.

8.3 Role of Soils in Plant Nutrition

Plants are dependent on soils for all of their nutrient elements other than C, H, and O. Even the H and O are derived from water that is taken up from soils (Figure 8.3). Plants intercept some water from precipitation, but that is only a minor part of the water used by most plants. Nitrogen, the fourth element in plants, is the most abundant element in the atmosphere, but it must be converted to NH_4^+ or NO_3^- by microorganisms in soils before it can be used by plants.

Green plants require nearly a score of nutrient elements, along with carbon dioxide and water. The most important of these, or the ones used in larger quantities, are N, S, P, Ca, Mg, K, and Cl, and all of theses are taken from the soil. Also, Fe, Mn, Zn, Cu, B, and Mo are essential elements for green plants, and a few other elements are

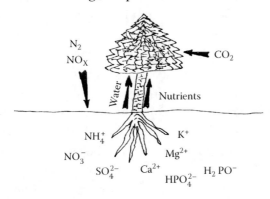

Figure 8.3 Plants take CO_2 in through stomata and acquire all nutrient elements other than C through their roots.

required by at least some plants. Most of the elements essential for green plants are critical for the organisms that decompose plants and incorporate the organic matter in soils also. Calcium and B are not important for the nutrition of fungi. Some plants require Si, mainly grasses, and Na is a major element in halophytes. Cobalt is an essential element in bacteria that fix nitrogen from N_2 in forms that are available to plants and animals.

8.3.1 Nitrogen

Nitrogen is a constituent of all amino acids and nucleotides, the building blocks of proteins and nucleic acids. These are the ingredients that are most definitive of life. Thus, N is not just an essential element, it, along with carbon, hydrogen, and oxygen, constitutes the foundation of living organisms.

Igneous rocks are practically devoid of nitrogen, and very little accumulates in sediments other than in organic matter within sediments. Most soil parent materials have only minor, or negligible, amounts of N in them. Plants require more N than any other element besides C, H, or O, and they benefit from extra N in practically any soil. Therefore, addition of N fertilizers is common agricultural practice. Until the Haber–Bausch process was developed in the twentieth century, the major source of nitrogen for fertilizer was the guano (bird excrement) and saltpeter ($NaNO_3$) that occur most abundantly in the arid climate of the Atacama Desert, along the Pacific coast of South America, where there are plenty of birds feeding on the abundant marine life of the Humboldt Current, and there is little precipitation to carry the soluble N-bearing bird waste products, and nitrates from other sources, away. Along with nitrate, phosphate and potash in the guano have enhanced its value as a fertilizer.

There is plenty of nitrogen in the atmosphere, which contains about a million times more than all of the living organisms on Earth, but it is not available to plants. Nitrogen from the atmosphere (N_2) must be converted to an available form before it can be taken up by plants. A large amount of energy is required to break the triple bond of dinitrogen (N_2). This can be done by lightning (Chapter 11), but most of the fixation of nitrogen in forms that are available to plants is accomplished by microorganisms.

Many different kinds of soil bacteria fix nitrogen. Where cyano-bacteria are abundant, as in wet and aquatic environments, they are much more effective in fixing nitrogen from N_2 than are non-photosynthetic free-living bacteria in soils (Maier 2000), especially where the cyanobacteria are associated symbiotically with water ferns (*Azolla* spp.). Symbiotic bacteria in root nodules on angiosperms are very effective in fixing nitrogen from N_2. They are such good suppli-ers of nitrogen that cultivated legumes are generally inoculated with *Rhizobium* sp. that live in nodules on roots of the plants and trade the nitrogen they fix from N_2 in available form for carbohydrates from the plants.

The life spans of bacteria are short, and they release inorganic and organic forms of N to the soil. Ammonia (NH_3) can be hydrolysed to ammonium (NH_4^-), and many different genera of chemotrophic and some heterotrophic bacteria oxidize NH_4^+ to NO_3^- (Maier 2000). Plants can then take up NH_4^+, NO_3^-, or monomeric forms of organic N, such as amino acids (Chapin et al. 2002; Schimel and Bennett 2004). Some plants harbor ericoid or ectomycorrhizal fungi that aid in the decomposition of organic polymers to monomers that can be acquired by plants (Bending and Read 1997). Otherwise, plants are completely dependent on free-living and symbiotic bacteria to make N available to them.

Calcifuges take up NH_4^+ preferentially and calcicoles take up NO_3^- preferentially (Marschner 1995). Plants have long been known to take up urea ($H_2N(CO)NH_2$) from fertilizers, but only recently has it been recognized that they can take up organic monomers (Schimel and Bennett 2004). Ammonium (NH_4^+) can be held on the cation-exchange sites of clay minerals and organic matter, but NO_3^- is read-ily leached from soils. There is so much demand for NO_3^- use by plants and microorganisms, however, that it generally has only a short residence time in soils before it is taken up by plants. Nitrogen loss by leaching may increase following a plant growing season when it becomes too cold for vigorous plant and microbial activity. In poorly drained soils with anaerobic conditions, more N may be lost as N_2O and N_2 gases than by leaching.

Plants cannot survive without free-living or symbiotic microorgan-isms to make nitrogen available to them. And animals cannot survive without plants to make amino acids for them.

8.3.2 Phosphorus

Phosphorus is a constituent in nucleic acids and a key element in the frameworks of DNA and RNA. Phosphates (PO_4^{3-}) have crucial roles in linking nucleotides in DNA and RNA, and in mediating chemical reactions. Nucleic acids, phospholipids, phosphorylated polysaccharides, phytin, and phosphoropyridine nucleotides are among the most important P compounds in plants (Stevenson and Cole 1999). Adenosine diphosphate (ADP) and adenosine triphosphate (ATP) are major conveyors of energy transfers within living organisms. The major role of ATP in energy storage and transfer within plants is mainly because of the energy release upon hydrolysis of the pyrophosphate bond in ATP (Marschner 1995). P deficiencies reduce plant growth and hinder maturation of seeds. After N, P is the chemical element most commonly limiting plant growth. A common feature of P deficiency is reddish leaf color caused by enhanced anthocyanin formation (Marschner 1995).

Apatite, or fluorapatite ($Ca_5(PO_4)_3F$) in which Cl^- and OH^- can substitute for F^-, is the main source of P in soil parent materials (Deer et al. 1966). Apatite is a common accessory in most igneous and metamorphic rocks and a minor constituent of sedimentary rocks. Phosphates have been concentrated in some sedimentary rocks and have been a major source of rock phosphate, which is commonly converted to fertilizer. The amounts of P in the igneous rocks represented in Table 8.1 are mainly proportional to the amounts of fluorapatite in them.

Phosphorus is released slowly by weathering of apatite. Once P is released by weathering, or accumulates from dust, only small amounts of it are lost by leaching from well-drained soils.

Most of the soil P that is not in minerals is in organic matter. It can accumulate as Ca-phosphate in alkaline soils and by adsorption on iron and aluminum oxyhydroxides in acidic soils. Phosphorus is most readily available in the pH 6 to 7 range. Organic P from plant remains and other organisms is decomposed throughout the growing season to supply plants with HPO_4^{2-} in neutral soils and $H_2PO_4^-$ in acidic soils (Stevenson and Cole 1999). In soils with low amounts of available P, mycorrhizal fungi are very helpful in obtaining P for plants.

As drained soils age over millions of years, adsorbed P is sequestered by immobile Fe and Al oxyhydroxides (Walker and Syers 1976)

Table 8.1 Chemical Elements, Other Than H, O, C, and N, Required by Plants or Other Organisms

Element		Ultramafic Igneous	Mafic Igneous	Intermediate Igneous	Silicic Igneous	Aluminous Sediment	Soils	Land Plants
Symbol	No.							
kg/Mg (parts per thousand)								
Na	11	5.7	19.4	30.0	27.7	6.6	12.0	0.2
Mg	12	259.0	45.0	21.8	5.6	13.4	9.0	0.7
K	19	0.3	8.3	23.0	33.4	22.8	15.0	3.0
Ca	20	7.0	67.2	46.5	15.8	25.3	24.0	5.0
Fe	26	98.5	85.6	58.5	27.0	33.3	26.0	0.5
g/Mg (parts per million)								
B	5	1	5	15	15	100	33	5
P	15	170	1400	1600	700	770	430	700
S	16	100	300	200	400	3000	1600	500
Cl	17	50	50	100	240	160	—	660
V	23	40	200	110	40	130	80	1
Cr	24	2000	200	50	25	100	54	<1
Mn	25	1500	2000	1200	600	670	550	400
Co	27	200	45	10	5	20	9	<1
Ni	28	2000	160	55	8	95	19	3
Cu	29	20	100	35	20	57	25	9
Zn	30	30	130	72	60	80	230	70

(*Continued*)

Table 8.1 (*Continued*) Chemical Elements, Other Than H, O, C, and N, Required by Plants or Other Organisms

Element Symbol	No.	Ultramafic Igneous	Mafic Igneous	Intermediate Igneous	Silicic Igneous	Aluminous Sediment	Soils	Land Plants
mg/Mg (parts per billion)								
Br	35	50	300	450	170	600	850	—
Mo	42	20	140	90	100	200	970	650
I	53	1	50	30	40	100	1200	—

Source: Data from Vinogradov, A.P., *Geochemistry, 7,* 641–664, 1962, for rocks; Shacklette, H.T., and J.G. Boerngen, *Element Concentrations in Soils and Other Surficial Materials of the Conterminous United States,* Professional Paper 1270, U.S. Geological Survey, Reston, VA, 1984, for soils; and Speidel, D.H., and A.F. Agnew, *The Natural Geochemistry of Our Environment,* Westview Press, Boulder, CO, 1982, for dried land plants.

Rock classes: ultramafic, dunite and peridotite; mafic, gabbro and basalt; intermediate, diorite and andesite; felsic, granite to granodiorite and rhyolite; aluminous sedimentary, clay or shale.

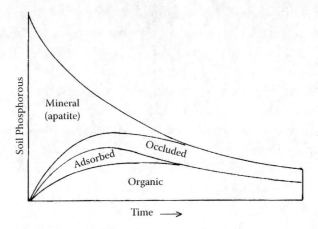

Figure 8.4 The relative magnitudes of four major compartments of soil P as they vary over millions of years.

such as goethite (FeOOH). The sequestered P is called occluded P (Figure 8.4). In very old tropical soils that lack soil organic matter and available P, some P is withdrawn from leaves before they fall (Vitousek et al. 1997).

8.3.3 Sulfur

Sulfur is a constituent of cysteine, cystine, and methionine and many proteins that incorporate cysteine and methionine (Paul and Clark 1996). Most of the sulfur in plants is in proteins and other organic compounds. Sulfur is crucial in linking complex polypeptide chains and maintaining folding patterns that are required to involve proteins in important chemical reactions. Sulfur is a key in many oxidation–reduction reactions in plants (Marschner 1995).

Sulfide contents of igneous and metamorphic rocks are low, especially in granitic rocks (Table 8.1). Volcanic emissions are sources of H_2S and SO_2. Sulfates accumulate in some sedimentary basins, and dust from the deposits is a source of S (for example, in gypsum, $Ca_2(SO)_4 \cdot 2H_2O$) for plants in arid environments. Some clayey sediments and shale contain appreciable amounts of S. Burning of coal, hydrocarbons, and biofuels is a major source of sulfur dioxide (SO_2). Sulfur dioxide in the atmosphere can be taken in by plants or oxidized to SO_3, which reacts with water to produce sulfuric acid, $H_2(SO)_4$.

Nearly all of the sulfur in most soils is in organic matter. Microorganisms decomposing organic matter recycle S through H_2S to SO_4^{2-} that can be taken up by plants. Sulfate that is not recycled back to organic matter or taken up by plants is readily leached from drained soils. Plants also acquire S by taking SO_2 in through their stomata.

Hydrogen sulfide is produced in some very poorly drained soils and in tidal marshes where the SO_4^{2-} from seawater is reduced to H_2S. Upon drainage of those soils, oxidation of the H_2S to SO_4^{2-} makes them extremely acidic. Those soils, called acidic sulfate soils, are major problems in some coastal areas.

8.3.4 Sodium and Potassium

These elements, Na and K, are present in rocks, soils, and plants as monovalent cations. They are regulators of osmotic potential and pH in plants. Potassium has many other functions. Sodium is essential in some desert plants, but it can interfere with the functions of K and be detrimental in most plants.

Silicic igneous rocks have more Na and K than mafic igneous rocks, and clayey sediments and shales can be important sources of them (Table 8.1). Most of the soil Na and K that is available to plants is held on the cation-exchange complexes of clay minerals and soil organic matter. The availability of K is more complicated, because it can be released from micas and vermiculite and become available to plants or be fixed between the silica layers of smectites to become unavailable. Deficiencies of K are common only in acidic soils from which K is readily leached out.

Sodium accumulates in poorly drained soil of dry climates. It is not only detrimental to most plants, but it can have profound physical effects on soils (Chapter 3).

8.3.5 Calcium and Magnesium

These elements, Ca and Mg, are present in rocks, soils, and plants as divalent cations. Calcium is an important component of cell walls and essential for growth of plants. It is an important conductor of signals within plants. Magnesium is a component of chlorophyll and an activator of many enzymes. It is essential for protein synthesis.

Mafic igneous rocks have much more Ca and Mg than silicic igneous rocks (Table 8.1), and the cations of these elements dominate on the exchange complexes of most soils. There is generally more exchangeable Ca than Mg. Ultramafic rocks, however, have so little Ca and so much Mg that only a minority of plant species can survive in soils derived from ultramafic rocks. Soils with Ca/Mg ratios < 0.7 mol/mol are genuine ultramafic soils (Alexander et al. 2007), commonly called serpentine soils. Limestone and dolomite are major sedimentary rocks that are composed of $CaCO_3$ (calcite) and $CaMg(CO_3)_2$ (dolomite). Pure limestone lacks most of the nutrients that are required by plants, but most limestone contains clay and silt impurities that are concentrated in limestone soils over hundreds or thousands of years as carbonated water dissolves the calcite and leaches the Ca from the soils.

$$CO_2 + H_2O = H^+ + HCO_3^-$$

$$CaCO_3 + 2 H^+ = Ca^{2+} + 2 HCO_3^-$$

Nutrient elements other than Ca are added to limestone soils in dust and precipitation. Plants that grow preferentially in limestone soils are called calcicoles, and those that avoid limestone soils are called calcifuges. Perhaps the most well-known calcicole plants are the cedars of Lebanon (*Cedrus lebani*). Vegetation on the flank of Lover's Leap (Figure 8.5) is typical of those on shallow limestone soils in the drier part (precipitation about 750 mm/year) of the Klamath Mountains in western North America.

8.3.6 Boron

Only minute amounts of B are present in soils. Plants acquire it from soil solution as H_3BO_3, or $B(OH)_3$, which is a very weak acid that is hydrolysed to H^+ and $B(OH)_4^-$ in plants. Although it is required by all plants, many of its functions in them are obscure. Most notably, it is important for the strength and development of cell walls and for plant growth by cell elongation.

Although there is little B in rocks, little is required by plants (Table 8.1). It can be leached from soils and cause plant deficiencies where the precipitation is great, or it can accumulate in arid and irrigated soils and become toxic to plants (Marshner 1995). The beneficial B,

Figure 8.5   The limestone landscape of Lover's Leap in the Klamath Mountains of western North America. The main plant communities are leatherleaf mountain mahogany/wheatgrass–cheat grass on shallow Mollisols, and ponderosa pine–western juniper/leatherleaf mountain mahogany/wheat-grass–cheatgrass on deep Inceptisols in colluvium. The plant taxonomic names, from larger trees to grasses, are *Pinus ponderosa, Juniperous occidentalis, Cercocarpus ledifolius, Elymus spicatus,* and *Bromus tectorum.*

or B tolerance range, is small but differs widely for cultivated plants. No data are given for plants in natural landscapes. Attiwill and Leeper (1987) report beneficial results from fertilization of Monterey pine (*P. radiata*) plantations with 8 kg/ha B.

8.3.7 *Chlorine*

There is little Cl in rocks, but precipitation generally supplies enough Cl to soils and plants that deficiencies are uncommon. The Cl⁻ ion is very soluble and highly mobile in water. The main functions of Cl in plants are charge compensation and osmoregulation (Marschner 1995).

8.3.8 *Transition Elements*

The transition elements are those between the first two columns on the left side of the periodic table of the chemical elements (Figure C.1) and the last six columns. There are three rows of transition elements, with 10 of them in each row, beginning with Sc in the first row, Y in the second, and Lu in the third. The biologically important transition elements are those from V to Zn in the first row and Mo in the second row.

All elements in the middle of the first row, from V to Cu, except Ni, have ions and compounds in which they occur with different valences. Because each of these elements has multiple valences, or oxidation states, the elements are important in oxidation-reduction reactions. In the reduction of nitrate to nitrite, for example, Fe and Mo supply the electrons to NO_3^- to reduce it to NO_2^-, $2\ Fe^{2+} + 4\ NO_3^- + 2\ H_2O = 2\ Fe^{3+} + 4\ NO_2^- + 2\ OH^- + H_2$.

Vanadium is required by fungi and marine algae, but apparently not by plants (Marschner 1995). It is an accessory element and not a major element in any minerals. The concentrations of V in soils are too low to be toxic to plants.

Chromium, although not an essential element, can be toxic to plants in the Cr(VI) oxidation state, as $HCrO_4^-$ or CrO_4^{2-}, but it is generally in the nontoxic Cr(III) oxidation state in soils. It is a beneficial element in animals (Ma and Hooda 2010). Chromium is most concentrated in ultramafic rocks, where it is present in chromite and other spinel group minerals. Chromium from industrial sources may be more of a threat to plants and animals than Cr(VI) occurring naturally in the environment.

Manganese is an activator of many enzymes and is a constituent of some proteins. It is involved in oxidation-reduction reactions. Manganese is prevalent in ferromagnesian silicates in which it substitutes for iron. In soils, Mn is present mainly in the Mn(II) and Mn(IV) oxidation states. Plants acquire it as Mn(II), which is most readily available in acidic soils. Plant toxicities can occur in acidic, waterlogged soils, which are prime conditions for Mn(II) mobility.

Iron is very important in oxidation-reduction reactions. Both Fe^{2+} and Fe^{3+} are involved in the catalysis of many reactions. In the porphyrin called *heme* it is a major transporter or electrons, as in the cytochromes of chloroplasts. Iron binds to oxygen in the heme protein called *leghemoglobin*, which is involved in nitrate reduction. Nonheme Fe-S protein, called ferredoxin, is very important in electron transfers in photosynthesis and other metabolic reactions. Iron is much more abundant than any other of the transition elements (Table 8.1). It is most abundant in Mg-silicate minerals, in which it substitutes for Mg, and in magnetite and other spinel group minerals. Some agricultural crops show Fe deficiencies in alkaline soils and toxicities in acidic soils (Welch et al. 1991). The deficiencies are caused by the immobility of Fe in alkaline soils, rather than low concentrations of Fe in the soils. Native plants that grow in soils to which they are adapted do not show Fe deficiencies of toxicities, but some eucalyptus trees that are adapted to acidic soils show Fe deficiencies when planted in alkaline soils (Attiwill and Leeper 1987).

Cobalt does not appear to be essential for plants, but is essential for microorganisms that fix N in legumes and other plants (Nicholas 1975). Cobalt substitutes for Mg in silicate minerals that are most abundant in serpentine soils. Cobalt is mobile in acidic soils, and is associated with Fe and Mn oxides. Some Co hyperaccumulating plants are reported in Africa, but none are found in North America.

Nickel is required in minute amounts by plants (Alexander et al. 2007). It is especially important in legumes where it is involved in urease metabolism (Eskew et al. 1983). Nickel is concentrated in ultramafic rocks (Table 8.1), where it substitutes for Mg and Fe in the silicate minerals. It persists in weathered serpentine soils as a component of clay minerals and with Fe-oxides and oxyhydroxides. Nickel occurs in soils only in the Ni(II) oxidation state, which is mobile in acidic soils. It can be toxic to plants, but it is more well known for its hyperaccumulation in plants (Baker et al. 2000). Because plants that are highly susceptible to Ni toxicity avoid serpentine soils, toxicities are generally found only in areas of mining or industrial pollution.

Copper is a micronutrient that is a constituent of several important enzymes in plants. Copper is most abundant in mafic rocks and clayey sediments (Table 8.1). The most common copper-bearing minerals are sulfides, such as chalcopyrite ($CuFeS_2$); it is not a constituent of silicate minerals (Norrish 1975). Copper is present in both monovalent and divalent oxidation states, most commonly as Cu(II). It is strongly adsorbed on Fe and Mn oxides in acidic soils, on clays, and on carbonates in alkaline soils. It is chelated with organic compounds and more concentrated than other plant micronutrient elements in organic soils (Shuman 1991), but it is bound so tightly that it is not readily available from soil organic matter. High soil organic matter concentrations can cause Cu deficiencies (Moraghan and Mascagni 1991). Copper deficiencies are most common in organic soils and soils in volcanic tephra. Soils contaminated by mining, industrial, munitions disposal, and agricultural sources of Cu generally remain contaminated for many years (Kabata-Pendias 2011). Hyperaccumulators of Cu have been reported, but these may be false reports caused by dust on the plants.

Zinc is a micronutrient with a variety of functions in plants. It occurs in a variety of ore minerals and substitutes for more common divalent elements in common rock forming minerals. Zinc is most abundant in mafic rocks and clayey sediments (Table 8.1). Deficiencies are common and widely distributed in agricultural crops, but not in natural plant communities. Zinc toxicities are present only in areas of Zn mining and processing. Several species of *Thlaspi* (a genus with many species, alternatively referred to as *Noccaea* spp.) that hyperaccumulate Ni also hyperaccumulate Zn (Reeves and Brooks 1983).

Molybdenum is the only second row transition element that is required by plants, and it is less concentrated in ultramafic rocks than in other kinds of rocks (Table 8.1). Although the quantities of Mo in soils are minute, very little Mo is required by plants. Molybdenum is an essential element in several enzymes. It occurs in the Mo(VI) and Mo(V) oxidation states and is particularly important for redox reactions in nitrate reductase and nitrogenase. Therefore, it is most important for plants that acquire N from soils as nitrates, rather than as ammonium ions. Tungsten, which is a third transition element, substitutes for Mo in some biochemical reactions in anaerobic microorganisms (Kabata-Pendias 2011). Molybdenum is most readily available to plants as $HMoO_4^-$ in neutral soils and as MoO_4^{2-} in slightly alkaline soils. Deficiencies, if there are any in natural plant communities, are most likely in acidic soils. On some poorly drained, generally alkaline soils, Mo in animals interferes with Cu metabolisim, causing molybdenosis (Welch et al. 1991).

Silver and *cadmium* in the second row, which are elements that are similar to Cu and Zn in the first row of transition elements, may accumulate in plants grown in anthropogenically contaminated soils, and Cd is then toxic to animals that consume the plants. Silver is not particularly toxic to animals, but it is toxic to some microorganisms and fungi (Kabata-Pendias 2011). Cadmium occurs as Cd^{2+} in soils and is highly mobile, but it is a hazard only for animals consuming plants grown in contaminated soils (Chaney 1977).

Tungsten, in the third row of transition elements, is chemically very similar to Mo. More than 2 billion years ago, before oxygen (O_2) became abundant, W may have been involved in some of the functions that have been assumed by Mo today (Williams and da Silva 1996). It is still used by some microorganisms.

Gold and *mercury*, in the third row of transitional elements, are present in soils only in minute amounts, and they have no roles in plant nutrition. Chemically, gold is not reactive and mercury is not very reactive, but Hg forms amalgams with several other metallic elements, including gold. The use of Hg in the concentration of Au in gold mining operations has been the largest source of mercury pollution. Mercury (Hg_o) vapor is highly toxic, and methyl mercury accumulates up the food chain in fish and in people whose diet includes much fish (for example, tuna) from contaminated water. Mercury is more common in the Hg(II) than in the Hg(I) oxidation state; it forms stronger covalent bonds than Zn or Cd (Pauling 1970), which facilitates the denaturing of proteins, which makes Hg so toxic in animals.

Selenium is required by few plants, but it accumulates to concentrations > 100 g/Mg in some plants. The chemistry of Se resembles that of S. Soils contain about 0.1 to 1 g Se/kg, with those derived from shale commonly having the greatest amounts (Kabata-Pendias 2011). Animals can show either deficiencies or toxicities to low or high Se concentrations. Plants growing in leached, acidic soils can have too little Se for animal health (Welch et al. 1991). Some plants (for example, *Astragalus* spp.) growing in areas with marine shales may accumulate toxic amounts of Se and be hazardous to grazing animals.

Bromine and iodine are not known to be essential for land plants. They are more closely associated with marine organisms and animals. Iodine is very important for animal nutrition.

Aluminum can be toxic to plants in extremely acidic soils (Pais and Jones 1997). Many elements besides Ag, Cd, and Hg, such as As and Pb, which are toxic to animals, are not important in plants.

8.4 Potential Plant Productivity

Having reviewed some of the drivers and controls of plant productivity, what are the patterns of ecosystem NPP across landscapes and around the globe? The main physical controls on plant productivity are climate

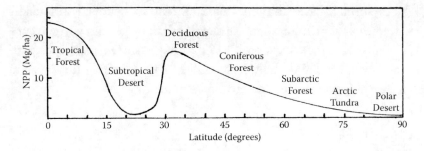

Figure 8.6 Net primary productivity of natural plant communities from the equator (0°) to the poles (90°). Only plant communities in ecosystems on an idealized continent with well-drained soils and lacking highlands and monsoonal climates are represented here. (From Leith, H., *Human Ecology*, 1, 303–332, 1973, and Atjay, G.L. et al., in B. Bolin et al. (eds.), *The Global Carbon Cycle*, Wiley, New York, 1979, pp. 129–181.)

and soils, including hydrology. Direct measurement of NPP is difficult and labor-intensive. With data from scores of sites, models were developed to estimate global NPP based on climatic parameters, mainly average temperatures and precipitation, or by computed evapotranspiration (Leith 1975; Atjay et al. 1979) (Figure 8.6). Subsequently, with data from hundreds of sites across the globe, Saugier et al. (2001) have summarized more refined estimates of NPP for eight biomes (Table 8.2).

Averaging NPP by biome obscures differences within biomes, which can be great. The Arctic nonforest (including tundra) in Table 8.2 is more a region than a biome. The annual NPP ranges from 0.001 kg/m² in polar deserts to about 0.2 kg/m² in mires and up to 1.0 kg/m² in

Table 8.2 Phytomass (Dry Matter) in Living Plants and NPP in Major Biomes of the World

Biome	Phytomass (kg/m²)				Annual NPP (kg/m²)[b]			
	Shoot	Root	Ratio[a]	Sum	Above	Below	Ratio[a]	Sum
Tropical forest	30.4	8.4	0.28	38.8	1.40	1.10	0.8	2.50
Tropical savanna	4.0	1.7	0.42	5.8	0.54	0.54	1.0	1.08
Deserts	0.35	0.35	1.00	0.7	0.15	0.40	2.7	0.25
Temperate grass	0.25	0.5	2.00	0.8	0.25	0.50	2.0	0.75
Temperate forest	21.0	5.7	0.27	26.7	0.95	0.60	0.6	1.55
Summer-dry scrub	6.0	6.0	1.00	12.0	0.50	0.50	1.0	1.00
Boreal forest	6.1	2.2	0.36	8.3	0.23	0.15	0.6	0.38
Arctic nonforest	0.25	0.4	1.60	0.6	0.08	0.10	1.2	0.18

Source: Data from Saugier, B. et al., in J. Roy et al. (eds.), *Terrestrial Global Productivity*, Academic Press, New York, 2001, pp. 543–557.

[a] Root:shoot ratio.
[b] Above and below ground NPP.

Arctic shrub (Shaver and Jonasson 2001). Besides climatic differences within each biome, soil water capacity, soil drainage, and nutrient supply can have large effects on NPP. Figure 8.6 shows landscapes with contrasting soils and plant communities having great differences in productivity within a temperate forest biome.

Questions

1. Lichens are an important component in the diets of some subspecies of reindeer, or caribou. Would you expect the NPP of those lichens, for example, those called reindeer lichen (*Cladina* spp.), to be very great?
2. Why do plant productivities decline when atmospheric temperatures are greater than about 30°C?
3. Will soil samples with $\delta^{13}C$ values of −26 most likely come from grassland or forest soils?
4. Which two of the following three factors are of major importance in determining the N status of soils, and which is of negligible importance?

   a. Soil parent material
   b. Soil microbial activity
   c. Plant productivity

5. Why is P such an important element in plants, even though its concentration is low in soil parent materials?
6. What is the fate of P in Oxisols and other very old soils?
7. What are the most common oxidation states of S in inorganic compounds, for example, in sulfides and sulfates?
8. Why are Ni toxicities not common in plants other than those in Ni-contaminated soils—soils contaminated by mining and industrial activities?
9. Is molybdenum (Mo) a more important element for plants that can acquire ammonium (NH_4^+) from soils or for plants that acquire N from soils as nitrate (NO_3^-) only?

Supplemental Reading

Bonan, G. 2008. *Ecological Climatology*. Cambridge University Press, Cambridge.

Chapin, F.S., III, P. Matson, and H. Mooney. 2002. *Principles of Terrestrial Ecosystem Ecology.* Springer, New York.

Etherington, J.R. 1982. *Environmental Plant Ecology.* Wiley, London.

Gurevitch, J., S.M. Scheiner, and G.A. Fox. 2002. *The Ecology of Plants.* Sinauer, MA.

Havlin, J.L., D.B.J. Tisdale, and W.L. Nelson. 1999. *Soil Fertility and Fertilizers.* Prentice-Hall, Upper Saddle River, NJ.

Jones, H.G. 1983. *Plants and Microclimate.* Cambridge University Press, Cambridge.

Kabata-Pendias, A. 2011. *Trace Elements in Soils and Plants.* CRC Press, Boca Raton, FL.

Leith, H. 1973. Primary production: terrestrial ecosystems. *Human Ecology* 1: 303–332.

Marschner, P. (ed.). 2012. *Marschner's Mineral Nutrition of Higher Plants.* Academic Press, London.

Marschner, P., and Z. Rengel (eds.). 2007. *Nutrient Cycling in Terrestrial Ecosystems.* Springer, Berlin.

Odum, E.P., and G.W. Barrett 2005. *Fundamentals of Ecology.* Thomson Brooks/Cole, Belmont, CA.

Whittaker, R.H. 1975. *Communities and Ecosystems.* Macmillan, New York. (A popular book with a good discussion of primary productivity)

9

SOIL ORGANISMS: LIFE IN SOILS

The products of photosynthesis in plants, algae, and some eubacteria are the basic food sources for terrestrial animals, fungi, bacteria, and eukaryotic microorganisms—both aboveground organisms and those in soils (Figure 9.1). Animals are the major feeders on plant materials aboveground, but smaller organisms, including many invertebrate animals, dominate the belowground activity (Mitchell 1988; Killham 1994; Coleman et al. 2004; Bardgett 2005; and a well-illustrated book by Nardi 2007). Living organisms are involved in practically all soil organic matter transformations. Earthworms, ants, termites, and larger animals mix organic detritus into soils, and redistribute organic matter within soils. Then soil microorganisms decompose the organic matter that has not been digested by the animals, or by microorganisms in the digestive tracts of the animals. Microorganisms are essential in breaking the organic matter down to components that can be reused by plants or other soil organisms, rather than allowing all of the organic matter to accumulate in soils. Enzymes of microorganisms are biological catalysts mediating soil organic matter reactions that would be very slow, or not even occur, in an abiotic environment.

The primary source of metabolic energy in soils is photosynthesis. Nonphotosynthetic organisms, other than autotrophic bacteria and algae, are dependent on green plants to fix carbon from the atmosphere. Most soil organisms feed directly on plants or plant detritus, or are indirectly dependent on plants by feeding on other living organisms or on the faeces of animals that feed on plant material or algae. They derive energy from the oxidation of organic compounds such as carbohydrates:

$$C_6H_{12}O_6 + 6O_2 \rightarrow 6CO_2 + 6H_2O + \text{energy}$$

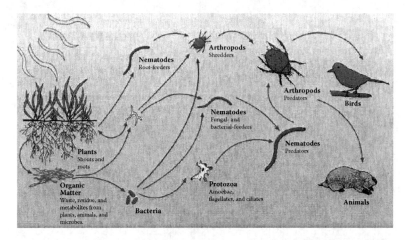

Figure 9.1   The food chain from organic matter produced by plants through bacteria and fungi, protozoa, nematodes, and arthropods (mites and insects) to chordate animals (birds and mammals). (From the United States Department of Agriculture Natural Resources Conservation Service.)

This is the reverse of photosynthesis (Chapter 8). A few bacteria derive energy from the oxidation of hydrogen and multivalent elements such as iron, nitrogen, and sulfur.

Vascular plants have green parts aboveground but roots in soils. Light from the sun seldom penetrates more than 1 or 2 cm below the ground surface. Photosynthesis is slight to nil below the surface. Thus, the soil is the domain of plant roots, microorganisms, and animals that do not derive energy directly from photosynthesis. Cavities, or pores, made by plant roots and animals are very important for the transmission of water and air, and thus the sustenance of life, in soils. Many animals living in soils lack eyes, or any visual organs, which are useless belowground where there is no light.

9.1 Faunal and Microorganism Habitats

Soils with different wetness, acidity, and nutrient status harbor different suites of organisms (Brown 1978; Dindal 1990; Madigan and Martinko 2006). The populations are greatest near the ground surface, where there is more organic matter and more aeration. Lichens, which contain cyanobacteria and microscopic algae, are limited to the immediate surface, because light for photosynthesis is greatly attenuated within millimeters from the ground surface and does not penetrate more than a centimeter or two into most soils. Microbial populations are generally high in the rhizosphere around plant roots.

Soils have cavities of all sizes, up to centimeters-wide burrows of rodents. Different organisms inhabit cavities of different sizes. Soil organisms range in size from bacteria that can inhabit nanopores in aggregates of clay particles, where they can evade all predators, to animals that make small to large tunnels and burrows (there are no standard definitions for pore size classes or organism size classes). The largest burrows are made by mammals.

Fungal hyphae reach into the larger nanopores, which are habitats of bacteria. Protozoa and nematodes inhabit micropores that retain water in soils at field capacity (water potential < −0.01 MPa). Root hairs can occupy micropores also. Mites and springtails are major occupants of mesopores, and earthworms are common inhabitants of macropores.

Many animals spend parts of there lives in soils and parts aboveground. Badgers and many kinds of rodents inhabit burrows in soils and forage aboveground. Many insects, such as some flies, enter soils to lay eggs, and their larvae live belowground until they become adult flies.

Some animals that are larger than the common soil pores push through soils by brute force (tunnellers), and others advance by digging (fossers) or by moving the soil bit by bit (miners). Some tunnellers are soft bodied and advance by peristaltic motion, with some ingestion of soil by earthworms, and others have rigid bodies (for example, millipedes). Moles, pocket gophers, and scarabaeid beetles are examples of fossorial animals, and ants and termites are common miners.

9.2 Viruses

Viruses are acellular constructs of DNA or RNA with a protein cover. They are < 100 nm long and commonly have tails up to 300 nm long. Viruses are parasitic within living organisms; reproduction of them occurs only within living organisms.

Viruses infect all kinds of living organisms, from bacteria to plants and animals. Those viruses that infect bacteria are called bacteriophages. Viruses are present in all kinds of environments, from desert and rainforest to Antarctica. The DNA or RNA of thousands of different kinds of viruses has been sequenced, and there may be millions more of them. Viruses are inactive in soils, outside of their hosts, and few persist very long in soils.

9.3 Main Kinds of Organisms Living in Soils

Millions of species of microorganisms and small animals spend all or part of their lives in soil (Richards 1987; Wood 1995). The belowground species diversity is generally greater than that aboveground. Many soil organisms have not yet been identified and classified by biologists.

Biologists group living organisms into two or three domains. The basic domains are eukaryotes (Eukarya), with chromosomes in a membrane-bound nucleus, and prokaryotes (Prokarya), lacking a membrane-bound nucleus. Prokarya are commonly split into two domains: Archaebacteria (or Archaea) and Eubacteria. Plants, Fungi, Protoctists, and Animals are four of an indefinite number of kingdoms within the Eukarya domain. Biologists have not agreed on the most appropriate number of kingdoms, or on how to distribute some of the organisms among the kingdoms (Tudge 2000) (Table 9.1, Appendix B).

A functional grouping of organisms is very helpful in defining the roles of organisms in the soil environment. Some organisms are free living and some are dependent on other organisms. Organisms are called *autotrophs* if they produce their own food or *heterotrophs* if they

Table 9.1   Six Kingdoms of Living Organisms: The Five Kingdoms of Whittaker with the Prokarya, that Whittaker called Monera, or Bacteria, Now Spanning Two Kingdoms, or Domains

	Kingdom	Kinds of Organisms (examples)
Prokarya		
	Archaea	Thermoacidophils, methanogens, and halophils
	Eubacteria	Proteobacteria, blue-green algae, actinomycetes
Eukarya		
No Embryonic Development		
Immobile	Fungi	Molds, mildews, mushrooms
Heterotrophic	Protoctists	Slime molds, protozoa
Photoautotrophic	Protoctists also	Diatoms, golden algae, yellow algae, green algae
Embryonic Development		
Photoautotrophic	Plants	Mosses, liverworts, vascular plants
Heterotrophic	Animals	Worms, slugs, insects, lizards, birds, mammals

Source: Whittaker, R. H., *Quarterly Review of Biology,* 34, 210–226, 1969.

Note: Later, Carl Woese separated the Prokarya into two domains, the Archaebacteria and the Eubacteria, with Eukaryota being the third domain (Woese and Fox 1977).

rely on other organisms, primarily plants, to produce organic compounds that the heterotrophs use as energy sources. Heterotrophs feed by absorption or ingestion of organic compounds or whole organisms (Table 9.1). Also, organisms may be grouped by size, habitat, food source, function, or whatever serves the purpose of the classifier.

Autotrophs that use solar energy to fix C are called *photoautotrophs*, and those whose primary source of energy is the oxidation or reduction of inorganic compounds are called *chemoautotrophs*. Green plants are photoautotrophs. All chemoautotrophs are prokaryotes (Archaea or Eubacteria).

Heterotrophs include predators that ingest and parasites that infect other living organisms and saprobes that live on dead organic matter. Organisms that live in or on other living organisms are *symbiotic*. The relationship is *mutualistic symbiosis* if both host and intruder benefit from it. The majority of soil organisms are heterotrophs rather than autotrophs.

Another grouping of soil organisims might be based on width, diameter, or other size parameters (Figure 9.2). Microorganisms the size of coarse clay or silt particles live in capillary pores, pores that retain water after a saturated soil drains. They are predominantly bacteria, actinomycetes, protozoa, and roundworms (nematodes). Mesofauna the size of sand grains and very fine pebbles occupy existing pores larger than capillary pores. They include potworms, mites, and springtails.

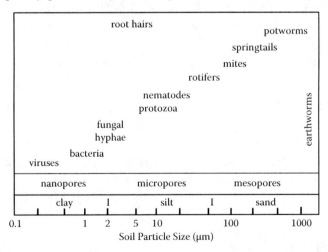

Figure 9.2 A size scale of soil organism width, or diameter, from viruses and microscopic bacteria to earthworms that are readily visible without magnification. The soil particle size scale is shown only for comparison to pore sizes—soils generally contain mixtures of sand, silt, and clay.

Macrofauna the size of pebbles and cobbles are large enough to make their own soil openings by digging (burrows) or force (tunnels). They include earthworms, mollusks, beetles, fly larvae, moles, rabbits, and rodents. Megafauna the size of boulders impact soils from the ground surface without entering them. Megafauna that impact soils include wild pigs, ungulate herds (for example, buffalo), and humans.

A habitat grouping of animals might focus on the parts of their lives that are spent in soils. Some animals live permanently in soils; some live in burrows (mammals) or tunnels (earthworms) and forage aboveground; some pass immature stages in soils and live aboveground as adults (flies and many other insects); and some that live aboveground impact soils by trampling (ungulate mammals) or routing (wild pigs).

Inasmuch as organisms in the six common kingdoms (Appendix B) have different roles in soil systems, a separate treatment follows for each kingdom. Although the Archaea, Eubacteria, Fungi, Animals, and Plants can be characterized by phylogeny, or genetic trends, Algae are differentiated from other Protoctists by their photoautotrophic inclination (Appendix B).

9.3.1 Archaea

Archaea lack a peptidoglycan layer that is characteristic of the cell walls of Eubacteria. These are pioneer microorganisms living in adverse environments that were more common billions of years ago, long before aerobic environments prevailed at the surface of Earth. They are represented by genera that survive in hot, saline (halophiles), or acidic environments. *Sulfolobus* spp. (thermoacidophiles) thrive in hot (70°C), acidic (pH 2) environments. *Methanobacterium* spp. (methanogens) live in anaerobic environments and produce methane (Table 9.2).

Soils lack suitable environments for most halophiles and thermophiles. Saline ponds and hot spring environments are more suitable for them. Methanogens occur in marshland and in the intestines and rumen of animals, where they live anaerobically and produce methane (CH_4).

9.3.2 Eubacteria

Eubacteria are represented by more individuals than all eukaryote kingdoms combined, yet their biomass in soils is relatively low due

Table 9.2   Genera of Bacteria Common in Soils or Performing Special Functions in Soils

Genera	Morphology[a]	Oxygen Requirement	Special Feature or Function
Archaebacteria (Ancient Bacteria)			
Methanobacterium	Rods, or cocci in chains	Anaerobes	Produce methane
Gracilicutes (Thin-Walled Eubacteria)			
Anabaena	Filamentous	Aerobes	Oxygenic photosynthesis[b]
Azotobacter	Rod, peritrichous flagella	Aerobes	Free-living N-fixers
Cytophaga	Rods	Aerobes	Gliding, nonfruiting bacteria
Desulfovibrio	Curved rods	Anaerobes	Reduce sulfur, N-fixers
Myxococcus	Indefinite, flexible cell wall	Aerobes	Heterotrophic gliding, fruiting bacteria
Nitrobacter	Short rods	Aerobes	Oxidize nitrite, reproduce by buds
Nitrosomonas	Rods, polar flagella	Aerobes	Oxidize ammonia to nitrite
Nostoc	Filamentous	Aerobes	Oxygenic photosynthesis[b]
Pseudomonas	Rods, polar flagella	Aerobes and facultative anaerobes	Reduce nitrate to N_2, reduce iron
Rhizobium	Rods	Aerosymbiotic	N-fixation in root nodules
Firmicutes (Thick-Walled Eubacteria)			
Actinomyces	Filaments	Aerobes	Produce conidiospores
Arthrobacter	Rods, spheres when young	Aerobes	Coryneform[c]
Bacillus	Rods, peritrichous or no flagella	Aerobes and facultative anaerobes	Form endospores, reduce manganese
Clostridium	Rods, peritrichous flagella	Anaerobes	N-fixers, form endospores
Frankia	Filaments	Aerobes	Symbiotic N-fixation
Streptomyces	Mycelium, multilocular sporangia	Aerobes	Produce actinospores
Thiobacillus	Rods, polar flagella	Aerobes	Oxidize S

Note: They are grouped by Archaebacteria and in two divisions of Eubacteria (Gracilicutes and Firmicutes).

[a] Polar flagella are at one end of rods, and peritrichous flagella are attached all around the perimeter of a cell.

[b] These O_2-producing photosynthesizers are blue-green bacteria that require light.

[c] *Coryneform* refers to club-like swellings on rods.

to the small sizes (0.3–1 μm) of individuals (Table 9.3). Nevertheless, the ecological impact of the bacteria is generally greater than that of any kind of other organisms, except plants. They comprise a metabolically diverse group that inhabits a very broad range of physical and chemical environments.

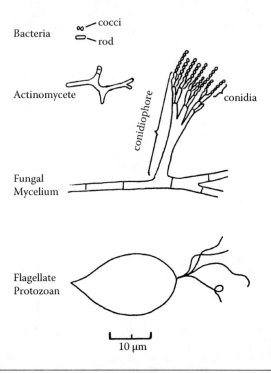

Figure 9.3 Some microorganisms that are common in soils, showing their relative sizes.

Three main divisions of Eubacteria are recognized by the nature of cell walls (Appendix B): Firmicutes that have thick cell walls, Gracilicutes that have thin cell walls with an outer liptoprotein layer, and Tenericutes that lack cell walls. Gracilicutes generally react negatively to Gram stain, and Firmicutes generally react positively. Tenericutes are not important in soils. Within each division, phyla and genera of bacteria are differentiated by morphology and metabolism.

Most bacteria in soils are rod shaped (Table 9.2), some are spherical (Figure 9.3), and some change shapes in different stages of growth. They commonly adhere to soil particles. Some bacteria require oxygen for growth; others do not. Those that require oxygen are called *(obligate) aerobes*. Those that require an oxygen-free soil atmosphere are called *(obligate) anaerobes*. Bacteria that do not require oxygen but can grow in the presence of O_2 are called *facultative anaerobes*. The production of enzymes that catalyse oxidation and reduction reactions in soils is a very important function of bacteria. A few specialized bacteria are responsible for the decomposition of large organic polymers in soils, while a great variety of less specialized bacteria are active in the further decomposition of diminished organic polymers and simple organic molecules. Many chemoautotrophic bacteria are involved in the transformation of inorganic compounds (Table 9.2). Bacterial populations vary manyfold in both time and space, depending largely on food supplies and other environmental conditions. Actinomyetes are effective in degrading chitin, cellulose, and other relatively decay-resistant polymers in neutral to alkaline soils.

Photosynthetic blue-green bacteria, or cyanobacteria, formerly called algae, live on or in surface soils to depths of 1 or 2 cm where light penetrates to those depths. Cyanobacteria are N-fixers, obtaining nitrogen, as well as carbon, from atmospheric gases. Some cyanobacteria live symbiotically with fungi and plants. Symbiotic colonies of fungi and cyanobacteria are called lichens. Cyanobacteria (*Anabaena*, *Lyngbya*, *Microcoleus*, *Nostoc*, *Scytonema*) and lichens are major components, along with bryophytes (mosses and liverworts), in soil crusts that form on soils with sparse vegetative cover.

Plants are dependent on prokaryotes for the fixation of nitrogen from the atmosphere, which is the primary source of soil N. Cyanobacteria and free-living eubacteria such as *Azotobacter* fix dinitrogen (N_2) as ammonium (NH_4^+), some (*Nitrosomonas*) oxidize NH_4^+ to NO_2^-, and others (*Nitrobacter*) oxidize NO_2^- to NO_3^- that is readily used by plants. Legumes have symbiotic relationships with protobacteria (*Rhizobia*) that form nodules on the plant roots and fix N that is available to the plants, and some shrubs and trees have symbiotic relationships with N-fixing actinomycetes (*Frankia*).

Actinomycetes are filamentous bacteria of about 0.5 to 2 μm in diameter that appear like miniature fungi (Figure 9.3). Other than the *Frankia*, actinomycetes are characteristically aerobic, free-living saprophytes. Their incomplete digestion of recalcitrant organic compounds produces geosmins that give soils an *earthy* smell. *Frankia* infect approximately 200 species of mostly woody plants in 24 genera. Alders infected with *Frankia* may fix 40 to 300 kg of N per hectare annually.

9.3.3 Protoctists

The majority of Protoctists, excluding most algae, are unicellular. They occupy 10 to 30 μm pores. Thus, they are more susceptible to drought than the bacteria that occupy smaller pores. Many Protoctists form cysts that are able to survive drought. Some Protoctists are parasitic on animals or have mutualistic relationships with them. Protoctists that are abundant in soils commonly feed on bacteria. They are concentrated upward in surface soils, because the bacteria that they feed on are concentrated there.

9.3.4 *Algae*

Algae are characteristically marine organisms, or in freshwater. Terrestrial algae, being photoautotrophic, are at the soil surface or aboveground on stones or plants. Diatoms and single-cell or filamentous green algae are found in some soil crusts. Soil crusts are common on soils with sparse vegetative cover, but the crust organisms are mostly mosses, liverworts, cyanobacteria, and lichens.

9.3.5 *Fungi*

Fungi are commonly the most abundant organisms in soils, when assessed by biomass, especially in acidic soils with plenty of organic matter. They grow by the production and extension of filaments called *hyphae*. A large mass of hyphae is called a *mycelium* (Figure 9.3). The number of individuals is difficult to ascertain, or perhaps meaningless (Table 9.3); therefore, the abundance is assessed by biomass (Table 9.4), rather than by the number of individuals. Fungi are abundant in moist soils and the dominant decomposers of organic detritus in very acidic soils.

Some fungi infect plant roots and aid the plants in obtaining water and nutrients from soils. Structures formed by fungal infection of plant roots are called *mycorrhiza*. Mycorrhiza with fungi whose hyphae enter plant cells are called *endomycorrhiza*, and those with fungi whose hyphae do not enter plant cells are called *ectomycorrhiza*, or sheathing mycorrhiza. Ectomycorrhiza are less common than endomycorrhiza, but they are very important for the survival and healthy growth of some trees, including pine trees. Ectomycorrhizal hyphae can extend several meters through soils—much further than endomycorrhizal hyphae can extend from plant roots.

Table 9.3 Common Relative Abundances of Different Kinds of Soil Organisms in Temperate, Nonforested Soils

Group	Number of Individuals (N) N/m^2	Approximate Biomass g/m^2
Eubacteria[a]	10^{13}	100
Actinomycetes	10^{12}	100
Protoctists	10^9	5
Fungi	—	150
Rotifers	10^5	0.1
Nematodes	10^6	2
Potworms	10^4	1
Earthworms	10^2	10
Mollusks	10^1	2
Arthropods		
Sowbugs	10^2	0.1
Mites	10^5	0.5
Centipedes	10^2	1
Millipedes	10^2	1
Pauropods	10^3	<0.1
Springtails	10^5	0.1
Beetles	10^3	0.1
Fly larvae	10^3	0.1

[a] Eubacteria other than actinomycetes.

Source: Data mostly from Wood, M., *Environmental Soil Biology*, Blackie Academic and Professional, London, 1995.

Table 9.4 Biomass (g/m^2, dry weight) of Major Groups of Organisms in Soils of Shortgrass Steppe in Colorado and in More Acid Woodland (Oak–Ash–Birch–Sycamore/Hazel Forest) in England

Group	Steppe	Woodland
Bacteria	60.1	3.7
Fungi	1.4	45.4
Protozoa	0.8	0.1
Nematodes	1.8	0.2

Source: Data from Wood, M. *Environmental Soil Biology*, Blackie Academic and Professional, London, 1995.

Note: The woodland plant biomass, including roots and litter on the forest floor, was 14.7 kg/m^2.

Arbuscular mycorrhizal fungi in the glomales order of the Glomero-mycota, formerly in the Zygomycota, produce a glycoprotein called *glomalin* that is very effective in enhancing the stability of soil aggregates.

9.3.6 Animals (Animalia)

This is a diverse kingdom (Figure 9.4, Table 9.5), from flatworms and microscopic rotifers to mammals, which are the largest animals in soils. Animals feed on plant roots, plant detritus, other animals, and animal wastes. They comminute plant detritus and animal remains to smaller sizes and mix them into soils where the material is more susceptible to microbial action.

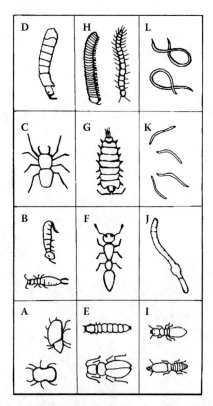

Figure 9.4   Major kinds of invertebrate animals in soils: (A) mite, (B) springtail, (C) spider, (D) fly larva, (E) beetle and larva, (F) ant, (G) woodlouse, (H) centipede and millipede, (I) termite, (J) earthworm, (K) potworms, and (L) nematodes. (From Klinka, K. et al., *Taxonomic Classification of Taxonomic Forms in Ecosystems of British Columbia*, Land Management Report 8, Ministry of Forests, Land Management, Victoria, BC, 1981; *Field Manual for Describing Terrestrial Ecosystems*, Land Management Report 25, Ministry of Environment Lands and Parks and Ministry of Forests, Victoria, BC, 1998.)

Table 9.5   Invertebrate Animals in the Upper 2.5 cm of
Two New York Forest Soils with Different Humus Types

Animal	Coarse Mull	Matted Mor
	Number/0.05 m^2	
Earthworms	18	0
Potworms	62	21
Mites	670	2182
Springtails	205	694
Bristletails	1	4
Pseudoscorpions	2	27
Spiders	5	8
Centipedes	0	4
Millipedes	17	0
Symphylids	4	23
Isopods (sowbugs)	43	0
Beetle larvae	11	38
Beetle pupae	4	3
Beetle adults	15	4
Fly larvae	12	13
Ants	1	49

Source: Data from Lutz, H.J., and R.F. Chandler, *Forest Soils*,
Wiley, New York, 1946.

Flatworms are the most primitive animals in soils, followed by rotifers and roundworms. Rotifers are common in some soils and sparse to absent in others. They live in soil pores containing water and feed on bacteria. Roundworms, or nematodes, are unsegmented worms that are abundant in soils.

Roundworms, or nematodes, are common parasites of plants and animals. Free-living, or nonparasitic, roundworms that live in soils are very small, with lengths < 1 mm and diameters equivalent to silt grains. They are present in all moist soils. The number of individuals is commonly about 1 billion per square meter ($10^9/m^2$) in the surface 10 cm of soil.

Roundworms are of substantial economic importance, because of their parasitism on plants and animals. Some nematode infections form galls on plants; some pierce plant roots, bury their anterior ends in them, and feed on fluids, or sap, from the roots; and others enter roots, leaves, or fruit and feed on tissue within plants or on fruit from plants. Free-living nematodes are active only in water, or films of water

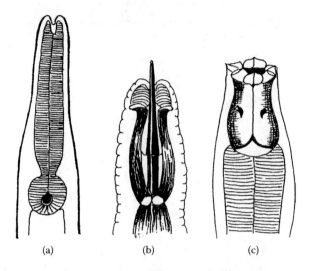

(a)   (b)   (c)

Figure 9.5 The mouthparts of nematodes with different feeding habits. (a) A saprophytic nematode. (b) A nematode parasitic on plants. (c) A predaceous nematode. (From Brown, A.L., *Ecology of Soil Organisms*, Heinemann Educational Books, London, 1978.)

in soils, but many survive dessication in inactive cyst-like form. They feed on microorganisms and fungi. A major role of roundworms in soils is the regulation of microbial populations. Nematodes with different feeding habits have different kinds of mouthparts (Figure 9.5). Saprophytic nematodes and bacterial feeders have narrow mouths and a long pharynx (Figure 9.5a); nematodes that are parasitic on plants have stylets that protrude from their mouths to pierce plant tissues (Figure 9.5b); and predacious nematodes are characterized by "teeth" around their mouths (Figure 9.5c).

Terrestrial mollusks are all pulmonate gastropods, slugs, and snails. Gastropod population densities of 10 to 25 per square meter are common in and on moist soils. Many gastropods feed on live plants or plant detritus at night and retreat to the soil during the day. Others feed on fungi, or on algae and lichens, and some are carnivorous, preying on earthworms or on other gastropods. Those that feed aboveground and retreat into moist soil incorporate organic matter into soils. Some gastropods feed on earthworms and seldom leave the soil.

Segmented worms, or oligochaetes, are common to abundant in most soils. The name *oligochaete* refers to sparse bristles, or lack of them. Most common in soils are the small potworms, or enchytraeids, of the mesofauna, and the larger earthworms (Figure 1.4) (Edwards and Bohlen 1996), or lumbriculids, of the megafauna.

Potworms, or microdriles, are generally about 1 to 5 mm long and white. The number of individuals ranges from about 10,000 to >100,000 per square meter in moist soil and under stones. They feed on bacteria, fungi, and possibly on partially decayed plant detritus. They are generally more prevalent than earthworms in very acidic soils. Though much smaller than earthworms, they have much greater metabolic rates that may make them as prominent in the processing of soil organic matter as are the earthworms.

Earthworms, or macrodriles, are reddish, and larger than potworms, but less numerous. Commonly, there are 10 to 200 lumbriculid worms per square meter in neutral moist soils. Some species of earthworms live in plant detritus, or litter, on the ground (epigeic earthworms); others inhabit subsoils to depths >1 m (endogeic earthworms); and those that inhabit vertical tunnels downward from ground surface are anecic earthworms. Epigeic and anecic earthworms are prolific tunnellers that are adapted to the low oxygen and high carbon dioxide contents of soils. Earthworms tunnel in friable soils by extending setae to anchor portions of their bodies and retracting setae in another portion and pushing the body ahead though the soil. These tunnelling movements are controlled by circular muscles that alter the coelomic fluid pressure in selected segments of the body. Earthworms tunnel in firm or clayey soils by ingesting soil though the mouth and ejecting it through the anus. This mixes soil and humus, and earthworms bring soil to the surface at annual rates of up to 8 kg/m^2. Stones have been covered by soil at rates of 3 to 5 mm per year. Earthworms are affected by drought, being inactivated after losing >10% of their water, but they survive water losses up to about 70%. They survive drought and frost by hibernation or estivation (dormancy). The beneficial effects of earthworms are the incorporation of plant detritus into soils and its partial digestion, mixing of soil by tunnelling, increasing soil aggregation, increasing soil drainage and aeration through their channels, and increasing soil water holding capacity and fertility. *Lumbricus rubellus* and *L. terrestris* mix plant detritus into soils by pulling leaves into their channels at night and feeding on them during the day. Ingestion of soil by earthworms mixes humus into it. Faecal material is less dense and holds much more water than ingested soil. Neutral

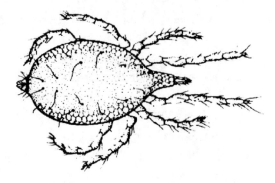

Figure 9.6   A soils mite, which is about the size of a sand grain. (From Andrews, W.A. (ed.), *A Guide to the Study of Soil Ecology*, Prentice Hall, Englewood Cliffs, NJ, 1973.)

soils are most favorable for earthworms, but some species inhabit acidic soils. They regulate pH by producing calcium carbonate. The faecal material commonly has more plant available nutrients than the ingested soil.

Arthropod species diversity is greater than that of any other group of animals in soils. They are animals without backbones (invertebrates), but with jointed legs. Among them, Arachnida (mites and spiders), Collembolla (spingtails), Myriapoda (millipedes and centipedes), and Insecta (insects and their larvae) are common in soils (Figure 9.4).

Mites are small arachnids that comprise one of the most abundant and ubiquitous groups of animals in soils. They are abundant in most soils, in cold and dry to hot and moist climates. Those that live in soils are small, about the size of sand grains (Figure 9.6). Mites feed on leaves, wood, bacteria, fungal hyphae, spores, and animal feces. Their role in soils is mainly the comminution of organic detritus, making the organic matter more susceptible to decomposition by microorganisms.

Two crustaceans have become well adapted to terrestrial environments: isopods (commonly called woodlice, sowbugs, or pillbugs) and crayfish. Woodlice are small bugs that are common in forest leaf litter. They crowd beneath stones, logs, or bark by day and emerge to feed on plant detritus or carrion at night. There are hundreds per square meter in favorable locations. Crayfish are abundant in some poorly drained soils with a fluctuating groundwater table. They burrow down to groundwater, sometimes when it is below 2 m, leaving a mounded ring of soil around each entrance of a burrow. Their burrows facilitate soil drainage and aeration, and burrowing and subsequent falling of detritus into the burrows result in soil mixing to depths > 1 m.

Myriapods are arthropods that have elongated bodies with many leg-bearing segments. Both centipedes having one pair of legs per body segment and millipedes having two pairs of legs per body segment are common in soils. Centipedes that live in the forest floor under stones and logs are commonly 3 to 6 cm long, and tropical ones are commonly about 20 cm long, but those that live belowground are generally smaller, and some in the order Geophilomorpha lack eyes. Most centipedes are poisonous and prey on other arthropods, earthworms, slugs, and nematodes. Millipedes range from 2 mm to about 25 cm long and have 11 to > 100 body segments. They live belowground and beneath stones, logs, bark, and leaves. Some inhabit earthworm channels, and some push through soil making their own tunnels. Most millipedes feed on plant detritus, some feed on live plants, and a few are carnivorous.

Symphylids, or garden centipedes, which resemble centipedes but have only 12 leg-bearing body segments, are common in soil and leaf litter. Those that live in soils are commonly 2 to 10 mm long. Most feed on plant detritus; some feed on live plant roots.

Pauropods are very small soft-bodied animals similar to millipedes but with 11 or 12 body segments and 9 or 10 pairs of legs. They are < 2 mm long. Pauropoda are commonly abundant in forest floor and inhabit soil also. They feed on fungal hyphae, plant detritus, and carrion.

Collembola (springtails) are the most numerous arthropods in many soils. They are very small, mesofaunal organisms < 6 mm long, with less total biomass in soils than some larger insects such as the Coleoptera (beetles). They live at different depths in soils and in plant detritus on the ground surface (Figure 9.7). Those that live aboveground have a linear appendage called a *furcula* near the rear of the abdomen that acts as a spring. The furcula is tucked under the body forward from its attachment on the abdomen and is extended rapidly to propel a springtail distances many times greater than its body length. Springtails that live more than a few centimeters below the ground surface have reduced springs or lack a furcula and eyes. Commonly there are about 10,000 springtails per square meter, many more than any insects. Springtails feed on bacteria, fungi, and plant detritus. Few are parasitic on plants and animals.

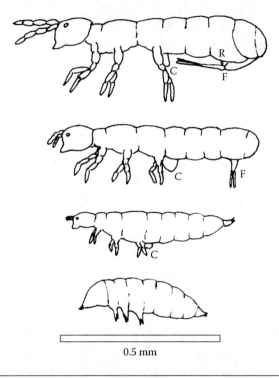

0.5 mm

Figure 9.7 Springtails that inhabit soils from the surface down to about 10 or 12 cm depth. Springtails at the ground surface (top) have a long furcula (F), a retinaculum (R) to hold the furcula under tension to be released for springing, and a well-developed collophore (ventral tube, C). All appendages, including legs, are smaller at depth, and springtails well below the ground surface are blind and white. (Redrawn partly from Nardi, J., *Life in the Soil*, University of Chicago Press, Chicago, 2007, and partly from Andrews, W.A. (ed.), *A Guide to the Study of Soil Ecology*, Prentice Hall, Englewood Cliffs, NJ, 1973.)

Winged insects generally have several stages of development that commonly have little resemblance to each other; for example, a fly has egg, larvae, and pupa stages that do not appear to be at all like an adult fly. Many winged insects spend one or more immature stages in soils and adult stages aboveground. This is true of beetles, possibly the most ubiquitous of the winged insects in soils, but some beetles are permanent soil dwellers. Hymenoptera (ants, bees, wasps) are common soil dwellers, but their distributions are more sporadic than those of beetles. Where ants and termites (Isoptera) are abundant in soils, they have much greater effects than beetles on the incorporation of organic matter and mixing of soils. Termites even have a major role in the decomposition of soil organic detritus, because they harbor cellulose-decomposing bacteria in their digestive systems. Termites are common in warm areas lacking severe winters. Ants and many bees and wasps live in colonies with highly developed social structures, but the bees and wasps living in soils are generally solitary. Some ants and termites burrow deeply and are the major movers of soil where there are few earthworms. Ants and termites are commonly evident from their mounds of soil and plant detritus. They generally feed on wood, other plant detritus, or fungi, and some ants prey on other arthropods.

There are four major groups of arthropods based on their feeding habits:

Shredders of mostly dead organic matter
    Macroshredders—millipedes and earthworms
    Microshredders—mites and springtails
Predators—ground (Carabidae) beetles, ants, centipedes, and pseudoscorpions
Herbivores (plant feeders)—root maggots and mole crickets
Fungivores (fungal feeders)—orbatid and endostigmate or astigmate mites, springtails

Many different species of chordates, mainly vertebrates, inhabit soils. A few salamanders, toads, turtles, lizards, snakes, and birds live or nest in soils, without much effect on them. Many mammals burrow extensively, and most of them have marsupial counterparts; for example, the wombat is a marsupial that is called a badger in Australia. Moles spend their entire lives in soils. They feed on earthworms and the larvae of insects in shallow burrows and nest in deeper burrows. Many rodents burrow extensively. Especially notable for their effects on soils, at least locally, are pocket gophers, which have been known to excavate as much as 18 tons of soil per hectare in a year, and prairie dogs (Figure 9.8). Large ungulate mammal herds, such as those of East Africa and the bison of prehistoric North America, must have had substantial trampling affects on soils.

9.3.7 Plants

Some prokaryotes obtain energy from the oxidation and reduction reactions of methane and N, S, Fe, and Mn compounds, but animals are dependent on plants to fix carbon in organic matter for their source of energy. Even though about one-half to three-quarters of the plant

Figure 9.8   A prairie dog community in Theodore Roosevelt National Memorial Park, North Dakota.

biomass is aboveground, that belowground is still much greater than the biomass of all other organisms in soils.

The dominant plants are vascular (Tracheata). They are able to lift water to great heights (100 m in some trees) and support many branches and leaves that gather solar energy and convert it to hydrocarbons that fuel the plants and organisms that feed on them and on detritus (stems and leaves) from the plants.

Nonvascular plants (Bryata) lack the vascular conductive tissues to support tall plants, and lacking roots and being smaller than most vascular plants, they convert much less solar energy to hydrocarbons. They are the dominant primary producers only on sites that are not amenable to vascular plants. Bryophytes are the primary producers on some very shallow soils and in climates that are too cold for vascular plants.

9.4 Soil Ecology

Soils are very complex systems, but they are not complete ecosystems. Most of the energy in a terrestrial ecosystem comes from plants that convert solar energy to stored energy aboveground. The stored energy is due to reduced carbon in organic matter. That organic matter produced by plants is the fuel for abundant life in soils. Primary production by autotrophic organisms in soils is negligible compared to that of plants that photosynthesize aboveground.

The distribution of organisms in soils, especially microorganisms, reflects the organic matter additions to the soils. Microorganisms are concentrated upward toward the surface in soils, in plant detritus on the surface, and around plant roots in soils. Most of the soil mesofauna are concentrated in these same places, because most of them feed on organic detritus or microorganisms. Larger fauna are more mobile and less concentrated about the primary food sources. Some megafauna feed aboveground and inhabit soils diurnally or irregularly.

The decay of organic matter is largely an organism-mediated process. It would proceed very slowly, or imperceptibly, if there were no life in soils. The amount, distribution, and kind of organic matter in an ecosystem is as dependent on the organisms in soils and in the plant detritus on the ground as it is on the plants that are the primary producers of organic matter. Soil mesofauna and megafauna comminute the plant detritus to smaller sizes and incorporate some of it into soils, where it is more accessible to microorganisms.

> Nutrients in dead leaves or needles are largely unavailable to most microbes. A bacterium in the leaf litter is analogous to a person in a pantry without a can opener.... The arthropod shredder is the bacterium's can opener. Bacteria and fungi will eventually use all of a dead leaf, but they are far more efficient if the leaf is shredded first. Shredders (i.e., millipedes, earthworms, sowbugs) crush vast quantities of plant cells from which they extract only the most readily available nutrients—the rest enter the normal soil recycling chain as the shredders defecate the crushed fragments. (Moldenke et al. 2000)

Only the more easily decomposed plant components such as simple carbohydrates, fats, and proteins are utilized by the larger organisms.

Microorganisms break any organic matter down chemically, incorporating some organic components into their body tissues, incorporating some into humus, and releasing some as inorganic compounds into soil solution or atmosphere. In warm, well-aerated soils, large quantities of organic matter do not accumulate. In cold or wet soils that are unfavorable environments for most soil organisms, organic matter does accumulate. The greatest prehistoric accumulations of organic matter have been in wet soils in warm climates where there has been plenty of plant production but too little oxygen in the saturated plant detritus on the ground for microbial activity to mediate its decay as rapidly as it accumulated. Coal is a lithified product of organic matter that accumulated in swamps many millions of years ago. The prime coal beds are on the order of 300 million years old. It was preserved in anaerobic environments to be oxidized by burning in modern furnaces.

Soil ecology is very important for the stability and sustenance of entire ecosystems. It must not be ignored, even though it is more difficult to observe and investigate than the aboveground components of ecosystems.

Questions

1. Plants are the main producers of organic matter in terrestrial ecosystems. Why is it important to have organisms to decompose those organic materials?
2. Does the variety and complexity of the organic matter in plant detritus require few or many different kinds of organisms for its decomposition?
3. Is the decomposition of plant detritus largely a competitive endeavor among different kinds of organisms, or are the activities of animals, fungi, and microorganisms more complementary in performing different processes in decomposition of the organic wastes from plants?
4. From a decomposer perspective, is it appropriate to call plant detritus waste?
5. The shredding of plant detritus is commonly a primary stage in the decomposition of plant detritus. Is that stage a more important preliminary operation for facilitating the activities of fungi or bacteria in decomposition processes?

6. How might soil compaction affect soil habitats and the distributions of soil organisms? Would it be more likely to curtail the activities of digger, tunnellers, or miners?

7. Can microorganisms escape from nematode predators in soil nanopores?

8. Different groups of invertebrate animals are known for different kinds of activities in soils. Name two of more classes of these soil animals that are in each of the following categories: macroshredders, microshredders, herbivores, fungivores, and predators.

9. In what parts of well-drained soils are microorganism and small invertebrate animal activities most concentrated?

10. Do you expect the ratios of heterotrophic to autotrophic microorganisms to be greater in surface soils or subsoils? Why?

11. What is the greatest limitation for microbial activity in continuously wet soils?

12. What features promote the accumulation of organic matter in soils?

Supplemental Reading

Andrews, W.A. (ed.). 1973. *A Guide to the Study of Soil Ecology*. Prentice Hall, Englewood Cliffs, NJ.

Bardgett, R. 2005. *The Biology of Soil*. Oxford University Press, New York.

Brown, A.L. 1978. *Biology of Soil Organisms*. Heinemann Educational Books, London.

Coleman, D.C., D.A. Crossley, and P.F. Hendrix. 2004. *Fundamentals of Soil Ecology*. Elsevier Academic Press, Amsterdam.

Dindal, D.L. 1990. *Soil Biology Guide*. Wiley, New York.

Killham, K. 1994. *Soil Ecology*. Cambridge University Press, Cambridge.

Madigan, M.T., and J.M. Martinko. 2006. *Brock Biology of Microorganisms*. Prentice Hall, Upper Saddle River, NJ.

Margulis, L., and M.J. Chapman. 2009. *Kingdoms and Domains*. Elsevier, Amsterdam.

Nardi, J. 2007. *Life in the Soil*. University of Chicago Press, Chicago.

Richards, B.N. 1987. *The Microbiology of Terrestrial Ecosystems*. Longman, New York.

Tudge, C. 2000. *The Variety of Life*. Oxford, New York.

Wood, M. 1995. *Environmental Soil Biology*. Blackie Academic and Professional, London.

10

SOIL ORGANIC MATTER

The origin of practically all of the organic matter in soils and terrestrial ecosystems is from green plants, although the organic matter of green plants is generally cycled through other kinds of organisms before becoming a component of soil organic matter (Chapters 8 and 9). Leaves falling to the ground accumulate in a litter layer of plant detritus that includes dead stems and branches. The organic tissues in this layer are shredded by invertebrate animals and gradually decomposed by microorganisms. Organic residues from the decomposition of plant stems, branches, and leaves accumulate at the ground surface, unless animals carry them into the soil. Earthworms, ants, and termites are notable among the animals that carry organic matter down in soils. Burrowing mammals are responsible for some mixing in soils, but generally much less than by arthropods (for example, insects, mites, centipedes, millipedes, and segmented (oligochaete) worms). Plant root decay is responsible for much of the organic matter that accumulates within soils.

Soils in different climates, and those on different kinds of landforms with different topography and relief, have different kinds of plant communities and different amounts of organic matter with different compositions. There is a great variety of soils with different forms of organic matter and different amounts and distributions of organic matter within the soils. Soil organic matter and its distribution within soils have large effects on soil properties that greatly affect the composition and productivity of the plant communities that they support.

10.1 Sources of Organic Matter and Its Composition in Plant Tissues

Plants are responsible for nearly all of the conversion of inorganic carbon to organic carbon in terrestrial ecosystems. The production of plant biomass increases from arctic to wet tropical biomes, whereas

Table 10.1   Biomass Production and Accumulation in Major Biomes of the World

Biome	NPP Annual	Plant Biomass	Distribution of Biomass			Litter	
			Wood	Root	Leaf[a]	Annual Input	Present
	Mg/ha (dry)		%			Mg/ha (dry)	
Tropical forest	10–35	60–800	74	18	8	30	5
Tropical savanna	2–20	2–150	60	28	12	10	3
Temperate grassland	2–15	2–50	0	83	17	8	5
Temperate forest	6–25	60–600	74	25	1	12	15
Boreal forest	4–20	60–100	71	22	7	8	35
Arctic forest	1–4	1–30	12	75	13	2	44

Source: Data from Swift, M.J. et al., *Decomposition in Terrestrial Ecosystems*, University of California Press, Berkeley, 1979, after Rodin, L.E., and N.I Basilevic, *Production and Mineral Cycling in Terrestrial Vegetation*, Oliver and Boyd, Edinburgh, 1967, and Whittaker, R.H., *Communities and Ecosystems*, Macmillan, New York, 1975.

[a] Leaf, may include other photosynthetic structures.

the rate of accumulation decreases (Table 10.1). Climates with greater plant productivity are also climates with more rapid rates of organic matter decomposition, except in wet soils where anaerobic conditions inhibit the decomposition soil organic matter. Consequently, the greatest accumulations of plant detritus are not in the tropics, but in colder regions that are not dry.

The composition of organic matter is essentially 1 part carbon and 1 part water with lesser amounts of N, P, and S, a few common cations, and minor amounts of many other chemical elements (Table 10.2). This is a gross simplification because the variety of organic compounds is practically endless. Carbon and N are much more concentrated in plants and soil organic matter than their averages in soils and the crust of Earth; P, S, and Cl are somewhat concentrated in plants; and all other of the 15 most abundant elements in plants are less concentrated in them than in soils and in the crust of Earth. Some N, P, S, and K is commonly withdrawn from leaves before they fall to the ground, reducing the concentrations of these elements in leaves added to the L (or O_i) horizon litter.

The organic tissues in plants are composed mainly of lipids (fatty acids, waxes, and oils), carbohydrates (either cold or hot water soluble), alkali-soluble hemicellulose, cellulose, lignin, and protein

Table 10.2 Amounts of Major Crustal Elements in Earth, Soil, and Plants

Element	Atomic Number	Plant Content (dry)		Soil g/kg	Earth Crust g/kg
		g/kg	mol/kmol		
C	6	440	280	16	0.2
H	1	65	490	47	1.4
O	8	460	220	—	460
N	7	13	7	1	0.02
S	16	1.5	0.4	1.2	0.3
P	15	1.1	0.3	0.3	1.0
Ca	20	8.5	1.6	9.2	36
K	19	6.1	1.2	15	26
Mg	12	1.4	0.4	4.4	21
Si	14	1.3	0.3	310	280
Cl	17	1.1	0.2	—	0.1
Na	11	0.5	0.2	5.9	28
Al	13	0.4	0.1	470	82
Mn	25	0.4	0.05	0.3	1.0
Fe	26	0.3	0.04	18	50
Ti	22	Trace	—	2.4	5.7

Source: Speidel, D.H., and A.F. Agnew, *The Natural Geochemistry of Our Environment*, Westview Press, Boulder, CO, 1982; Shacklette, H.T., and J.G. Boerngen, *Element Concentrations in Soils and Other Surficial Materials of the Conterminous United States*, Professional Paper 1270, U.S. Geological Survey, Reston, VA, 1984; Clarke, F.W., *The Data of Geochemistry*, Bulletin 770, U.S. Geological Survey, Reston, VA, 1926.

(Table 10.3). Lignin, cellulose, and hemicellulose are the major components of plant cell walls, which are the parts of plants that are most resistant to decomposition. Lignin and tannins are polyphenols that are very resistant to decomposition and greatly influence the rates of plant tissue decay. Lignin is composed of an indefinite network of alcoholic aromatic benzene rings with no fixed chemical formula (Tate 1989; Stevenson 1994; Paul and Clark 1996). There are three principle kinds of the monomer building blocks in lignin (Figure 10.1). Only the ρ-coumaryl alcohol units are present in monocotyledonous angiosperms; sinapyl alcohol units are found in both dicotyledonous and monocotyledonous angiosperms; and coniferyl alcohol units are found in both gymnosperms and angiosperms. Cellulose is another complex constituent of plant tissues that is moderately resistant to

Table 10.3   Composition of Organic Materials That Are Potential Sources for Soil Organic Matter

Material	Lipid	Carbo-hydrate[a]	Hemi-cellulose	Cellulose	Lignin	Protein	Ash
Wood							
Conifer	3–10	2–8	13–17	48–55	23–30	—	<1
Eudicot	2–6	1–2	19–24	45–48	17–26	—	1
Pine needle	24	7	19	16	23	2	2
Oak Leaf							
Young	8	22	13	16	21	9	6
Old	4	15	16	18	30	3	5
Hairgrass[b]	2	13	24	33	14	2	—
Cow	50	?	0	0	0	39	11
Arthropods	13–26	14–31	5–9	—	0	38–50	?
Earthworms	2–27	11–17	—	—	0	54–72	9–23
Fungi	1–42	8–60	2–15	—	0	14–52	5–12
Bacteria	10–35	5–30	4–32	—	0	50–60	5–15

Source: Data from Swift, M.J. et al., *Decomposition in Terrestrial Ecosystems*, University of California Press, Berkeley, 1979; obtained from many different sources.

Note: Percentage of dry weight.

[a]   Soluble and stored carbohydrates, including chitin in fungi and arthropods.

[b]   *Deschampsia flexuosa.*

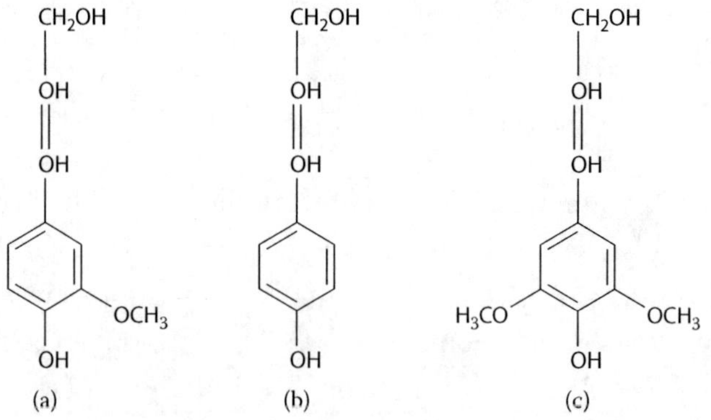

Figure 10.1   Building blocks of lignin. (a) Guaiacyl or coniferyl alcohol. (b) ρ-Coumaryl alcohol. (c) Sinapyl alcohol.

decomposition. It consists of chains of glucose units that are hundreds to thousands of units long, with hydrogen bonds between adjacent chains uniting them to produce strong fibers such as those in cotton. In contrast to lignin, cellulose has a more definite chemical formula, which is $(C_6H_{10}O_5)_n$, although the number of glucose units (n) is indefinite. Hemicellulose, which is less resistant to decomposition than cellulose, consists of short, branched chains of sugar units. Besides glucose, xylose is a common sugar unit in hemicellulose, and mannose, galactose, rhamnose, and arabinose are others. Gelatinous pectin and fibrous xylans are common hemicelluloses.

10.2 Decomposition of Organic Matter

The decomposition of plant tissues is a biological process (Berg and Laskowski 2006). It requires water; even dry rot will not occur in an atmosphere completely devoid of moisture. The resistance of plant tissues to decomposition is dependent largely on the compositions of lignin, cellulose, and hemicellulose in their cell walls. Microorganisms are crucial for decomposing these cell constituents. Animals do not produce enzymes that are capable of attacking lignin, and their capabilities of digesting cellulose are limited, but they have important roles in shredding plant issues containing lignin and cellulose and making these components and other constituents of plant tissues more accessible to microorganisms. Some animals, however, harbor bacteria that allow them to digest lignin and cellulose; two common examples are termites that can digest lignin and ungulate mammals that digest cellulose in their rumen. The termites have very strongly alkaline fluids in their guts and harbor streptomycetes.

Only fungi, mainly basidiomycetes, and very few bacteria produce enzymes that degrade lignin. Lignin is not degraded completely by any one enzyme, but in increments by lignin peroxidase, manganese peroxidase, and other enzymes. Lignin and cellulose molecules are large, and their breakdown must be initiated externally by enzymes of fungi and filamentous bacteria (actinomycetes) before fragments of them can be incorporated into the cells of bacteria for more complete decomposition. Fungi that decompose wood fall into two broad categories, which are called white rot and brown rot. White rot decomposes lignin and leaves white cellulose. Brown rot decomposes cellulose and hemicellulose, leaving lignin that is brown upon oxidation.

Litter decomposition is greatest in warm, moist climates and very limited in extremely cold and dry climates. After climate, the lignin content of plant tissues is the best indicator of their resistance to decomposition (Meentemeyer 1978; Aerts 1997). Following lignin, N concentration is helpful in predicting rates of plant tissue decay.

Phospholipids, proteins, nucleic acids, and organic phosphates are rapidly decomposed in plant litter and soils, but some amino-N associated with lignin or with the clays in soils is somewhat resistant to decomposition.

Rates of organic matter decomposition (Box 10.1) are dependent not only on climate and plant tissue composition, but also on soil wetness. Oxidation is very important in the decomposition of plant tissues and soil organic matter. Organic matter accumulates under anaerobic conditions in wet soils, as indicated by the high content of organic C in Histosols, which are predominantly organic matter.

10.3 Composition of Soil Organic Matter

The initial products and residues of microbial activity and plant decomposition (carbohydrates, proteins, amino acids, lipids, and lignins) are transitory. They are decomposed further or transformed by microorganisms and polymerized to form complex organic substances that are relatively resistant to decomposition. Collectively, the more resistant products are called *humus*. Humus compounds are brown to dark brown or black amorphous polymers of high, but variable, molecular weight (MW about 1000 to 100,000 g/mol). They range from small linear to large spherical molecules. A single molecule, if spherical, would have a diameter in the 0.5 to 2 nm range, which is about the thickness of a phyllosilicate clay layer. (Inorganic primary soil clay particles, however, generally contain many more than one layer of phyllosilicate, and are therefore much larger.) Waksman (1932) proposed that the humus in soils represents modified lignins, but the formation of humus is now recognized to be much more complex (Stevenson 1994).

Although the composition of humus is indefinite, chemical fractionation has been applied to separate three or more kinds of humus. The most basic units are (1) humins that are insoluble in a dilute NaOH solution (pH 9–10), (2) fulvic acids that are soluble in extremely acidic solution, and (3) humic acids that are not soluble in extremely acidic solution (Figure 10.2). Sizes of the humus polymers increase from about 1 to 30 kDa in fulvic acids and 10 to 100 kDa in humic acids (Stevenson 1994; Paul and Clark 1996) (Da is the symbol for Daltons, or g/mol).

BOX 10.1: RATES OF SOIL ORGANIC MATTER DECOMPOSITION

Organic matter decay rates follow first-order reaction kinetics. The loss of organic matter (Y) with time (t) is proportional to its concentration (Y):

$$-\delta Y/\delta t = kY$$

where k is a constant.

$$Y/Y_0 = e^{-kt}$$

where t_0 is the concentration of Y at time zero.

Substituting 1 for Y_0 and 1/2 for Y and taking the natural logarithms to find the half-life of the organic matter ($t_{1/2}$),

$$\ln 0.5 = -k\, t_{1/2}$$

$$-t_{1/2} = 0.693/k$$

Jenkinson (1977) analyzed organic matter data from long-term experiments in England and proposed that the data represent five organic matter (OM) "compartments" with half-lives ranging from 0.165 to 1980 years:

Decomposable plant material: 0.165 years
Resistant plant material: 2.31 years
Soil (microbial) biomass: 1.69 years
Physically stabilized OM: 49.5 years
Chemically stabilized OM: 1980 years

Jenkinson (1997) found that the radiocarbon (^{14}C) age of carbon in the soil of the long-term experiments was 1450 years in the upper 23 cm, 2000 years in the second 23 cm, and 3700 years in the next 23 cm, below 46 cm. Most of the organic matter in the soil has been there for a long time.

Another way to assess the rate of soil organic matter decay is by what is called the turnover time. At steady state, when the amount of plant material added to a soil each year (A) is equal to the amount of organic matter lost from the soil, A = kY, and rate of turnover (Y/A) is equal to 1/k; for example, if the amount of material added in a year is one-tenth of the amount in the soil, Y/A = 10 and the turnover time, 1/k, is 10 years.

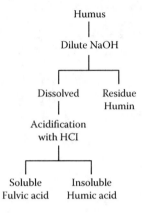

Figure 10.2 Fractionation of soil organic matter to differentiate humin, fulvic acid, and humic acid.

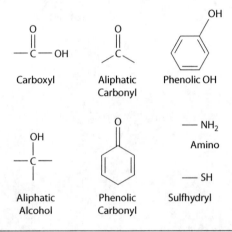

Figure 10.3 The main functional units in humus.

Humus complexes have large surface areas and a variety of reactive functional units. The most common functional groups (Figure 10.3), in order of decreasing abundance, are (1) carboxyl, (2) phenolic OH, (3) allophatic (alcoholic) OH, (4) quinoid carboxyl, (5) ketonic carboxyl, (6) amino, and (7) sulfhydryl units. Humus has many potential cation-exchange sites, depending on the soil pH, and relatively few anion exchange sites. Cation exchange, as indicated by the titratable acidity (Table 10.4), is attributable to (1) carboxylic acid, (2) phenolic OH, and (3) allophatic OH units. The cation-exchange capacity (CEC) is most closely related to the titratable acidity of the carboxylic acid from which H^+ dissociates at relatively low pH; H^+

Table 10.4 Characteristic Elemental Compositions of Fulvic and Humic Acids in Soils and the Acidities of Major Functional Units in Them

Acid	Elements (g/kg)						Acidity by Unit (mmol+/kg)[a]		
	C	H	O	N	S	Ash	HOCH$_2$	COOH	Phenolic OH
Fulvic	495	45	449	8	3	24	500	9100	3300
Humic	564	55	329	41	11	9	1000	4500	2100
Humin	575	—	330	—	—	—	—	3500	<2000

Source: Data from Haider, K., *Soil Biochemistry*, 7, 55–94, 1992, after Paul, E.A., and F.E. Clark, *Soil Microbiology and Biochemistry*, Academic Press, San Diego, 1996; data for humin from Filep, G., *Soil Chemistry* (English trans.), Akadémiai Kaidó, Budapest, 1999.

[a] Most of the titratable acidity is attributable to three functional units. The phenolic OH is attached to an aromatic (benzene) ring.

dissociates from the alcoholic OH units at higher pH values than those considered in CEC determinations.

As the phytomass in detritus, or plant litter, decomposes, organic carbon is lost, decreasing the C:N, C:S, and C:P ratios. These ratios decrease from live plants, through leaf litter, to humus in inorganic soils (Table 10.5) as C is lost to the atmosphere in CO_2 and other gases and by the leaching of bicarbonates from soils. The C:N ratios generally decrease to < 12 in subsoils, although there are exceptions in soils with very acidic leaching and translocation of organic acids downward. The C:N ratios of fungi are in the range from 15:1 to 4.5:1, and the ratios in bacteria are about 5:1 to 3:1 (Paul and Clark 1999).

Table 10.5 Typical Organic C to Organic N, P, and S Ratios (g/g)

	C:N	C:P	C:S
Plants			
Mean (Table 3.1)	34	400	290
Litter on Ground			
Conifer forest	40	320	240
Broadleaf forest	30	270	600
Surface Soil Humus			
Conifer forest	25	—	150
Broadleaf forest	20	160	140
Grassland	11	80	90

Note: Organic C:N ratios are <12 in the subsoils of practically all soils that have mull type organic matter distributions.

The complete oxidation of organic matter yields carbon dioxide. Under anaerobic (lack of oxygen) conditions that exist in very poorly drained soils and many waste dumps, methane (CH_4) is an end product of decomposition.

The recycling of soil P through plants and microorganisms is so efficient that generally little is lost from soils by leaching. Therefore, as organic carbon is lost during detrital organic matter decomposition, the C:P ratio decreases. It decreases to < 200 (Table 10.5) and then increases in intensively weathered and leached soils as P is gradually lost over many thousands of years.

The carbon-to-sulfur ratios are on the order of 200 in fresh plant detritus and 100 in humus. Reduced sulfur compounds, such as hydrogen sulfide (H_2S), may accumulate in some poorly drained soils, and pyrite (Fe_2S) may form by precipitation of the sulfide.

10.4 Distributions of Soil Organic Matter

The amount of organic matter that accumulates in soils depends on the balance between production and decay (Amundson 2001). Both plant production and organic matter decay increase with higher temperature in well-drained soils. Because the rate of decay is greatly affected by temperature, warm tropical soils generally have less organic matter than cold soils (Figure 10.4, Table 10.6). This idealized concept of the latitudinal distribution of soil organic matter is complicated

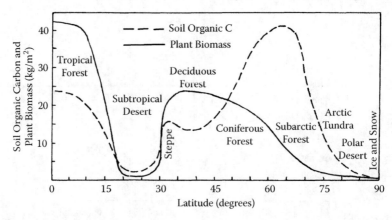

Figure 10.4   Latitudinal plant biomass and soil organic matter trends in well-drained soils. Soil organic matter content is approximately double that of soil organic carbon.

Table 10.6 Organic Carbon in Plant Litter and Soils of the Orders in *Soil Taxonomy*

Order	Organic Carbon (kg/m^2)	
	Above Ground	Below Ground
Alfisols	8	6.9
Andisols	Variable	30.6
Aridisols	<1	3.5
Entisols	Variable	9.9
Histosols	—	204.6
Inceptisols	Variable	16.3
Mollisols	2	13.1
Oxisols	2	10.1
Spodosols	18	14.6
Ultisols	5	9.3
Vertisols	1	5.8

Source: Belowground data from Eswaran, H. et al., *Soil Science Society of America Journal*, 57, 192–194, 1993, and organic C in above ground plant litter estimated from the generalization of data in Table 10.1 and data in Van Breeman, N., and P. Buurman, *Soil Formation*, Kluwer, Dordrecht, The Netherlands, 1998. Eswaran et al. (1993) have no data for Gelisols.

by the relationships of precipitation to latitude, and by soil drainage. Plant production and soil organic matter content are low around the subtropical high atmospheric pressure zones at about 20 to 30° north and south latitudes, because the air descending in these high pressure zones is too dry to yield enough precipitation to support much plant growth. Precipitation is low in polar zones also.

Well-drained polar soils have little organic matter, because they are relatively dry and unproductive. Poorly drained polar soils, however, have much organic matter. Soils with permafrost (permanent ice) are generally poorly drained, with the exception of those that are too dry for the formation of continuous layers of ice. The permafrost beneath polar soils stores large quantities of organic matter. Organic matter accumulates in many wet soils in all climatic zones, because saturation retards the decay of organic matter. Discontinuously wet soils with very thick surface layers containing more than about 26% organic matter and continuously wet soils with more than 16 to 24% organic matter in sandy to clayey materials are called *organic soils* (Soil Survey Staff 1999, where percentages are for organic carbon, rather than for

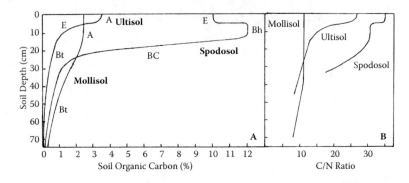

Figure 10.5 Soil depth distributions of carbon in soil organic matter and the organic C:N ratios in a neutral grassland Mollisol, a moderately acidic deciduous forest Ultisol, and a very strongly acidic conifer forest Spodosol.

organic matter). Organic soils occur at all latitudes. In the United States, Alaska has the largest area of wetlands and organic soils, followed by Minnesota and Florida. California has a large area of wetlands and organic soils in the delta of the Sacramento and San Joaquin Rivers.

Soil organic matter is concentrated nearer the surface in boreal forest soils, particularly in cool conifer forests with relatively little animal activity, than in grassland (steppe in Figure 10.5) (recall that soil organic matter (SOM) = 2 × soil organic carbon [SOC] soils). About one-half or more of the grass biomass is belowground, compared to about one-quarter to one-third of the tree biomass in conifer and broadleaf forests. The forest floor of plant litter, moss, and lichens contains much of the soil organic matter in boreal forests. (Soil analyses generally exaggerate the amounts of organic carbon in the E horizons of boreal forest soils, because the E horizons are commonly thin, and it is difficult to sample them without incorporating some material from adjacent horizons that contain much more organic matter.) A substantial portion of soil organic matter in grassland, or steppe, soils is derived from fine root decay. There are large areas of steppes at about 30 to 35° latitude (Figure 10.5) on the southern Great Plains of North America and on the Pampas of South America, as might be expected from global patterns of atmospheric circulation, but there are larger areas of steppe at higher latitudes on the northern Great Plains of North America and in Central Asia.

The C:N ratios are much higher in the A horizons of forest soils and in subsoils of cold or wet Spodosols than in grassland soils represented by Mollisols (Figure 10.5). The C:N ratios converge to about 8 to 10 or 12 in the lower subsoils of all of these soils.

10.5 Functions of Soil Organic Matter

Soil organic matter is very important for its effects on soil structure (next chapter) and fertility. The importance to fertility is due to (1) the storage of plant nutrients and their release for plant use, especially N, P, and S, when organic matter decays, and (2) the cation retention capacity of soil organic matter. The CEC of most soil organic matter is in the range of 1.5 to 3 mol+/kg (150–300 meq/100 g, or 1500 to 3000 mmol+/kg) at pH 7 to 8. It is greater in neutral to slightly alkaline soils than in acidic soils, because the negative charges responsible for cation retention are due to the dissociation of organic acids, which increases as the soil pH increases. The surface area of humus is about 800 to 900 m^2/g. Thus, soil organic matter with its large surface area and high CEC has pronounced effects on the physical activity of soils.

Humus and polysaccharides bind to the clays in soils by a variety of mechanisms and promote aggregation of soil particles. Some of the mechanisms are illustrated in Figure 10.6. Negative charges predominate on both clay minerals and the major components of

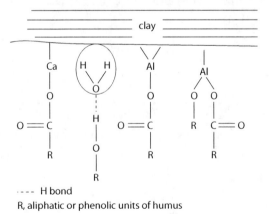

Figure 10.6   Some of the links that bind humus to clay minerals. The Ca represents divalent cations (Ca, Mg, Fe, and Mn), and the Al represents trivalent cations (Al and Fe).

humus. Divalent and trivalent cations act as intermediaries to bind the clays together, along with the attractions of hydrogen bonds, the surface tension of water molecules, and van der Waal's forces. The R components in Figure 10.6 may represent polysaccharides, lipids, or proteins, or they could all be components of one humus molecule or particle. Aggregation of soil particles by organic matter reduces their susceptibility to detachment by wind and water and their dispersion and displacement in water. Organic soils that are dominated by organic matter, however, are highly susceptible to wind erosion when they are dry.

A major effect of soil organic matter is to increase soil porosity and aeration. For example, an average California soil composed of quartz and feldspar grains, clay minerals, and no organic matter would have a mean particle density of 2.6 Mg/m^3, a bulk density of 1.66 Mg/m^3 (Alexander and Poff 1985), and 36% porosity, whereas a similar soil with 9.6% organic mater (4.6% organic carbon) would have a mean particle density of 2.5 Mg/m^3, a bulk density of 1.0 Mg/m^3, and 58% porosity. Thus, increased organic matter in mineral soils (organic matter < 20 or 30%) has minor effects on mean particle density, but major effects on bulk density and soil porosity.

Greater organic matter reduces soil strength, which, along with increased aggregation of soil particles, facilitates the penetration by plant roots both between peds and within them. Thus, increased organic matter makes soils more favorable physical environments for plant roots, and also for animals.

Problems

10.1. A soil has 22% clay that has a mean CEC of 300 mmol+/kg and 5% organic matter. What is the approximate CEC of the soil based on A, the low, and B, the high range of soil organic matter CEC?

Answer:
a. $0.22 \times 300 + 0.05 \times 1500 = 66 + 75 = 135$ mmol+/kg
b. $0.22 \times 300 + 0.05 \times 3000 = 66 + 150 = 216$ mmol+/kg

10.2. A soil at steady state (concentration of organic matter constant from year to year) contains 18 kg C/m^2, and a mean of 1.8 kg/m² of plant detritus (0.9 kg C/m^2) is added to the soil each year. What is the turnover time (1/k) of the soil organic matter?

Answer: A = kC, 0.9 = 18 k, k = 1/20, 1/k = 20 years

10.3. Nearly all of the atoms in fulvic and humic acids are C, H, and O atoms, with a few more C than H or O atoms. Assuming that the ratios of C:H:O atoms are 1:2:1, how many C atoms are there in 30 kDa molecules of fulvic or humic acids? The proportion of C atoms is proportional to its gram atomic weigh in CH_2O, 12/(12 + 2 + 16) = 0.40. Then the amount of C in a mole of the acid is 0.4 × 30,000 g/mol = 12,000 g C/mol. Since the atomic weight of C is 12 g, the number of atoms in a mole of the acid is 12,000/12 = 1000 C atoms per molecule of the acid.

Questions

1. In which biome of Table 10.1 is the turnover rate of plant detritus (litter) greatest?
2. In which biome of Table 10.1 is the turnover rate of plant detritus (litter) slowest?
3. Why are 11 of the 12 most abundant chemical elements in plants among the first 20 elements, those with atomic numbers no greater than 20?
4. Based on the lignin contents of wood and grass (hairgrass) in Table 10.3, would you expect conifer wood, pine needles, or grass to be decomposed more rapidly?
5. Why are divalent and trivalent cations more effective than monovalent cations in binding soil organic matter to clays and promoting soil aggregation?
6. Notice in Table 10.6 that most of the organic matter of Mollisol and Oxisols is belowground (for different reasons) and most of the organic matter of Spodosols is aboveground. Can you explain these differences?

7. List some ways that soil organic matter increases soil resistance to erosion.

8. List some soil organic matter influences that make soils better media for plant growth.

Supplemental Reading

Paul, E.A., and F.E. Clark. 1996. *Soil Microbiology and Biochemistry*. Academic Press, San Diego.

Speidel, D.H., and A.F. Agnew. 1982. *The Natural Geochemistry of Our Environment*. Westover Press, Boulder CO.

Stevenson, F.J. 1994. *Humus Chemistry, Genesis, Compositions, Reactions*. Wiley, New York.

Stevenson, F.J., and M.A. Cole. 1999. *Cycles of Soil: Carbon, Nitrogen, Phosphorus, and Micronutrients*. Wiley, New York.

Swift, M.J., O.W. Heal, and J.M. Anderson. 1979. *Decomposition in Terrestrial Ecosystems*. University of California Press, Berkeley.

Tate, R.L. 1989. *Soil Organic Matter*. Wiley, New York.

11

SOILS AND GLOBAL PROCESSES

Soils are open systems. They are not isolated from the rest of the world. Matter and energy are exchanged between soils and surrounding environments. Soils are always changing in response to their changing surroundings. Fluctuations in soil heat and water were discussed in Chapters 4 and 5. The water cycle will be revisited and mass transfers will be discussed on a global scale in this chapter. Mass transfers are discussed both as total mass and as individual chemical elements.

Two of the most important cycles are water and carbon. These are essential for the sustenance of living organisms. The maintenance of temperatures in the 0 to 100°C range in which water is liquid makes our planet, Earth, a unique habitat for life. Nitrogen and sulfur were abundant in the environments where life first developed on Earth, and they are major constituents in living organisms. There is more to it than abundance, however, because less abundant phosphorus is also a major element in living organisms. In order to place the cycles of these major chemical elements in a global perspective, it is helpful to begin with a discussion of the outer layer of Earth that geologists call the crust. Although crustal development and cycling involves many millions (or billions) of years, it is the current cycling of water and chemical elements over much shorter intervals that is of more general interest. This cycling involves the outer margin of Earth's crust, the oceans, and the atmosphere. It is the upper surface of the crust where soils are located and plants and other organisms live.

11.1 Crust of Earth

Soon (millions of years) after Earth was formed about 4.55 billion years ago, the materials in it separated into three distinct layers (Figure 11.1). At the centre is a *core* of iron alloy. Around the core is a viscous *mantle* that is essentially magnesium silicate with abundant

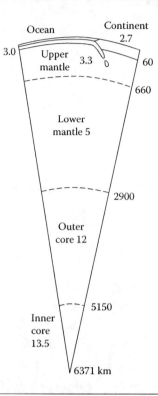

Figure 11.1   A slice of Earth. Depths are distances (km) from the surface of Earth. Numbers from 2.7 in upper continental crust to 13.5 in the inner core are densities (mg/m³).

iron, substantial aluminum and calcium, and appreciable sodium, titanium, chromium, manganese, and nickel (Table 11.1). The mantle is 2900 km thick. At the top of the mantle is a rigid, solid crust that ranges from 6 to 12 km thick in ocean basins to 30 to 75 km thick on continents. Above the crust, and water bodies on it, is the atmosphere where nitrogen and oxygen are abundant.

Ocean crust forms from magma and lava derived by partial melting of the uppermost mantle. It is mostly gabbro and basalt and includes sediments that accumulate in ocean basins.

This ocean crust has relatively high concentrations of calcium, magnesium, iron, and titanium compared to continental crust (Figure 11.2), although the concentration of magnesium is much less than in the mantle. Large plates of ocean crust drift to subduction zones where most of the ocean crust sinks back into the mantle. No ocean crust on the plates is older than about 210 Ma, indicating geologically rapid cycling of ocean crust back to the mantle. Most of the

Table 11.1   Concentrations of First 30 ($z = 1 - 30$) Chemical Elements in Living Organisms, the Lithosphere, and the Sea—Means or Medians

Element Z and Symbol	Upper Mantle[a]	Ocean Crust[b]	Continental Crust Bulk[c]	Continental Crust Upper[c]	Soils[d] (U.S.)	Ocean[a] mass/L	Land Plants[e]	Land Animals[e]
			mg/kg			mass/L	mg/kg	mg/kg
1 H	—	—	—	—	—	—	67 g/kg	74 g/kg
2 He	—	—	—	—	—	—	—	—
3 Li	2	9	13	20	24	0.2 mg	2	<1
4 Be	<1	—	2	3	1	0.2 ng	<1	<1
5 B	<1	—	10	15	33	4.5 µg	25	1
6 C	150	140	—	300	25,000	28.0 mg	463 g/kg	510 g/kg
7 N	10	—	—	—	—	0.4 mg	19 g/kg	98 g/kg
8 O	—	—	—	—	—	—	396 g/kg	268 g/kg
9 F	16	—	—	500	430	1.3 mg	4	>150
10 Ne	—	—	—	—	—	—	—	—
11 Na	2600	20,000	23,000	28,900	12,000	10.8 g	1200	—
12 Mg	222,000	46,000	32,000	13,000	9000	1.3 g	3200	1000
13 Al	22,000	81,000	84,000	80,000	72,000	0.3 mg	200	—
14 Si	210,000	236,000	268,000	308,000	310,000	2.5 mg	3000	>120
15 P	80	700	—	650	430	65 µg	2000	>17000
16 S	180	800	—	—	1600	0.9 g	4800	5000
17 Cl	1	—	—	100	—	18.8 g	2000	2800
18 Ar	—	—	—	—	—	—	—	—
19 K	250	880	9100	28,000	15,000	0.39 g	11,000	7400
20 Ca	26,000	81,000	53,000	30,000	24,000	0.45 g	15,000	>20
21 Sc	17	38	30	11	9	0.9 ng	<1	<1
22 Ti	1100	9700	5400	3000	29,000	10 ng	32	<1
23 V	82	290	230	60	80	2.2 µg	2	<1
24 Cr	3200	300	185	35	54	1.2 ng	2	<1
25 Mn	1000	1000	1400	600	550	72 ng	240	<1
26 Fe	64,000	81,600	71,000	35,000	26,000	0.2 µg	200	160
27 Co	100	47	29	10	9	1.2 ng	1	<1
28 Ni	2000	150	105	20	19	0.5 µg	2	1
29 Cu	28	74	75	25	25	0.2 µg	10	2
30 Zn	50	100	80	190	230	0.3 µg	50	160

Note: Concentrations for Ca, P, F, and Si in some animals are minima—they can be much higher in the skeletal parts of those animals.

[a] Li, Y.-H., *A Compendium of Geochemistry*, Princeton University Press, Princeton, NJ, 2000; seawater concentrations in nanograms (10^{-9} g) per liter, unless other units are specified.

[b] Hoffman, K.A., *American Scientist*, 256, 96–103, 1988; ocean ridge basalt.

[c] Taylor, S.A., *Solar System Evolution*, Cambridge University Press, Cambridge, 1992.

[d] Shacklette, H.T., and J.G. Boerngen, *Element Concentrations in Soils and Other Surficial Materials of the Conterminous United States*, Professional Paper 1270, U.S. Geological Survey, Reston, VA, 1984.

[e] Romankevich, E.A., in V. Ittekkat et al. (eds.), *Facets of Modern Biogeochemistry*, Springer-Verlag, Berlin, 1990, pp. 39–51, concentration in dried specimens.

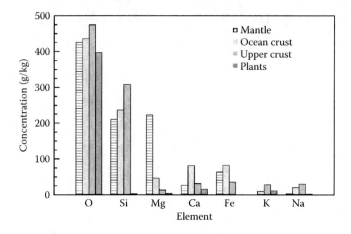

Figure 11.2 Major chemical element concentrations in the upper mantle, ocean crust, upper continental crust, and terrestrial plants.

ocean crust is subducted to great depths in the mantle (Figure 11.1), but some is pushed onto a continental margin or into a subduction trench, and some is plastered beneath the outer edge of a continent as the ocean crust descends in the process of subduction. Slowly continental crust is produced, and some is cycled back to the mantle. The net rate of continental crust accretion is controversial (Condie 1997). Currently continental crust covers about 38% of Earth's surface.

Plants and other living organisms live at the surface of the upper continental crust, and soils reside there. This upper crust is highly differentiated topographically, lithologically, mineralogically, and chemically. Some of the differentiation is caused by tectonic and volcanic processes, and some is caused by other processes that are operating at the surface of Earth. Continental crust rises higher than ocean crust, because it is thicker and lighter, making it float higher over the mantle. Most of the upper surface of continental crust is above sea level.

A brief summary of the natural processes that are operative on the continents includes the influences of the atmosphere, wind, water, ice, and gravity. Water flows from the continental landscapes and collects in continental and ocean basins. Rocks exposed on continental surfaces are subjected to the corrosive and erosive actions of water and the atmosphere, which is called *weathering*. Weathering of rocks and soils and erosion yield particulate and dissolved products that are carried away in runoff and deposited along streams and in sedimentary basins.

In the processes of fluvial and lacustrine transport and deposition, gravel and sand are commonly separated from silt and clay. Chemical and biological processes add to the complexities of soils and sedimentary deposits by concentrating salts, silica, lime, or organic matter in many of them.

Sediments that are deeply buried are transformed by consolidation to conglomerate, sandstone, and shale, and if subjected to high enough pressure and temperature, they are transformed by metamorphism to gneiss, schist, phyllite, or slate. The chemical and biological sediments are transformed by consolidations to chert, limestone, or lignite, and by metamorphism to quartzite, marble, or coal. Salt deposits, such as gypsum (calcium sulfate) and halite (sodium chloride), are not extensive. Consolidated sedimentary deposits and their metamorphic heirs are commonly returned to the surface, after millions of years of burial, by tectonic uplift and are subjected to another cycle of weathering, erosion, and sedimentation. Many rocks of continental platforms, such as the Canadian Shield, have survived for more than a billion years without deep burial, but exposures of them have been weathered and they have contributed sediments to the oceans. Also, volcanic areas, which can cover thousands of kilometers (for example, about 160,000 km^2 on the Columbia Plateau of western North America and 2 million km^2 on the Karoo Plateau of southern Africa), are contributing sediments to the crustal surface cycle.

11.2 Free Water Balance

The origin of Earth's water is a mystery. Was it present when Earth was formed in our stellar environment, or did Earth accumulate water later as it was bombarded by comets and other stellar bodies during the Hadean era? We know that oceans must have been present by the Archean era, about 3.8 Ga ago, to support the growth of cyanobacteria, which have been called blue-green algae. Some water is bound in the mantle and in igneous rocks, some is bound in sedimentary rocks (*connate water*), and some is free to circulate through the regolith and above.

Free water is considered to be that which is not bound in rocks or minerals. It is sometimes referred to as *meteoric* water, water that is free to cycle through the atmosphere and regolith. The amount of

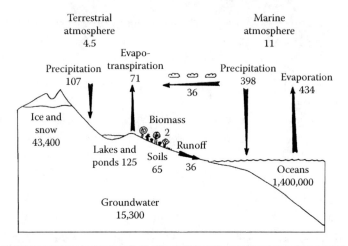

Figure 11.3 The global water cycle: reservoirs and fluxes, or transfer of water between reservoirs. The units are 10^3 km^3 (thousands of cubic kilometers) or 10^{12} Mg (or tons), or 10^{18} g. (Data from National Research Council, *Global Change in the Geosphere-Biosphere*, National Academy Press, Washington, DC, 1986.)

meteoric water has been practically constant for millions (or billions) of years. Relatively little meteoric water is currently added in tectonic and volcanic activities, and relatively little escapes by diffusion from Earth. Some water is carried back into the mantle of Earth by subduction of crustal plates, and some is added to the meteoric water supply when comets fall to Earth.

Most of the meteoric water (97%) is in oceans, about 2% in ice caps and glaciers, and 1% in groundwater systems, including minor amounts in soils and streams (Figure 11.3). Although only about 0.001% of the meteoric water is in Earth's atmosphere, it plays a major role in cycling water evaporated from oceans, and water transpired from plants, back to the ground in precipitation. Dry air descending through the troposphere at about 20 to 30° north and south latitudes picks moisture up from the oceans and carries it poleward in westerly winds and toward the equator in tropical convergent winds. The result is greater precipitation near the equator and at about 40 to 60° north and south latitudes (Figure 5.5).

Storage of water is limited in terrestrial ecosystems, yet enough water is cycled through them to make them much more productive than marine ecosystems. Even though water is the main component of living organisms, they contain only a minute fraction of the meteoric water.

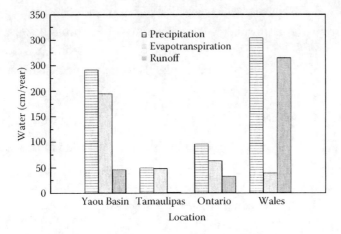

Figure 11.4 Precipitation–evapotranspiration-runoff patterns from the Yaou Basin, French Guiana, 4° latitude; Tamaulipas, Mexico, 24° latitude; Ontario, Canada, 44° latitude; and Wales, United Kingdom, 53° latitude.

Water enters terrestrial ecosystems primarily from the atmosphere, by precipitation. Some ecosystems receive water that flows in from adjacent ecosystems also. Water is lost from ecosystems by overland flow (aboveground), through the ground to streams or to groundwater tables, and by evaporation and transpiration from soils and plants. The combination of evaporation and transpiration from terrestrial ecosystems is called *evapotranspiration*. It is closely related to air temperature, the vegetative cover, and the amount of solar radiation at the ground or canopy level (Chapter 5). In dry climates, most of the scanty precipitation is returned to the atmosphere by evapotranspiration. In wet climates, although much water may be lost by evapotranspiration, more of it flows through soils to streams or groundwater tables (Figure 11.4). Water losses by overland flow, over the ground surface rather than through soils, are generally not great in undisturbed ecosystems. Most water from precipitation infiltrates into soils on its way to plant roots, groundwater tables, or streams. Overland flow, however, can be substantial with intense rainfall or rapidly melting snow in disturbed ecosystems. Some ecosystem altering disturbances that allow greater overland flow of water, such as wildfires, are common natural phenomena.

Important aspects of the water cycle are the transfer of heat from one place to another and the transport of suspended particles and dissolved substances. Water washes particles and dissolved constituents from the atmosphere; it carries substances through soils, it is a major agent of erosion and particle loss from soils, and it is the main conduit of mass transport from uplands to sedimentary basins.

11.3 Carbon Cycle

Carbon, hydrogen, and oxygen are three of the four most abundant elements in our solar system, and they comprise the bulk of organic tissues. Life on Earth is dependent on a favorable carbon cycle.

On Earth, C is concentrated in the continental crust and in the oceans (Figure 11.5), particularly in the oceans (Table 11.2), mainly as Ca and Mg carbonates in limestone and dolomite and as carbonate

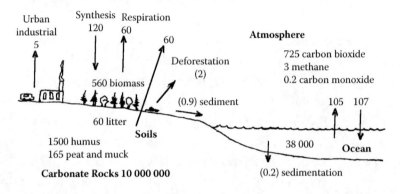

Figure 11.5   The global carbon balance. Numbers represent petagrams (10^{15}), or gigatons (Gt), of carbon in the compartments and annual fluxes (petagram/year). (Adapted from Schlesinger, W.H., *Biogeochemistry: An Analysis of Global Change*, Academic Press, San Diego, 1997, and other sources.)

Table 11.2   Near Surface Stores of Carbon, Nitrogen, Sulfur, and Phosphorus on Earth

Reservoir	Carbon	Nitrogen	Sulfur	Phosphorus
	Gt (or Pg, 10^{15} g)			
Atmosphere	750	3.8×10^6	3.6×10^3	30×10^{-6} (dust)
Oceans	38×10^3	540	1.2×10^6	88
Ocean photic zone	1000			
Living organisms	560	23	6	2
Soils	1500	95	94	26

Source: Data from Schlesinger, W.H., *Biogeochemistry: An Analysis of Global Change*, Academic Press, San Diego, 1997, and several other sources.
Note: Ocean volume assumed to be 1.35×10^9 km^3, 1.35×10^{18} m^3.

and bicarbonate ions in seawater. Concentration of C in the upper mantle is on the order of 150 g/Mg (ppm) (Li 2000), and volcanism is only a minor source of C, practically negligible. Carbon dioxide, which currently comprises about 0.04% of the gas in the atmosphere, is the primary source of carbon for plants. Carbon is actively cycled between the atmosphere and oceans and between the atmosphere and plants, with soils being major repositories for C fixed in reduced forms by plants (Table 11.3). The decomposition of the organic matter in soils releases carbon dioxide (CO_2) from well-drained sites and methane (CH_4) from wet sites with anaerobic soils.

Currently, atmospheric CO_2 and CH_4 are major climatic concerns. Earth receives short-wave solar radiation from the sun and emits long-wave radiation (Chapter 4). Carbon dioxide, methane, and other atmospheric gases with larger molecules than N_2 and O_2 absorb

Table 11.3 Organic C in Living Plants and Soils and NPP in Biomes of the World

Biome	Area 10[6] km²	Plant Carbon[a] Total Pg	km/m²	Soil Carbon[b] Total Pg	kg/m³	Annual NPP (C) Total Pg	kg/m²
Tropical forest	17.5	340	19.4	201	11.5	21.9	1.25
Tropical savanna	27.6	79	2.9	149	5.4	14.9	0.54
Deserts	27.7	10	0.4	108	3.9	3.6	0.13
Temperate grassland	15.0	6	0.4	199	13.3	5.6	0.37
Temperate forest	10.4	139	13.4	82	8.1	8.1	0.78
Summer-dry scrub	2.8	17	6.1	21	7.6	1.4	0.50
Boreal forest	13.7	57	4.2	224	16.4	2.6	0.19
Arctic shrub	1.5	1	1.0	3	2.2	0.3	0.22
Arctic sedge-shrub	0.9	2	1.7	13	14.5	0.1	0.11
Arctic sedge mire	1.0	1	0.5	18	18.2	0.1	0.10
Arctic semidesert	1.4	<1	0.1	2	1.3	<0.1	0.02
Arctic desert	0.8	<1	<0.1	<1	<0.1	<0.1	<0.01
Cropland	13.1	4	0.3	—	—	—	—
Ice	15.5	0	0.0	0	0.00	0.0	0.00
Sum	148.9	656	—	1020	—	58.6	—

Source: Data from Saugier, B. et al., in J. Roy et al. (eds.), *Terrestrial Global Productivity*, Academic Press, New York, 2001, pp. 543–557; Shaver, G., and S. Jonasson, in J. Roy et al., *Terrestrial Global Productivity*, Academic Press, New York, 2001, pp. 204–210; and Post, W.M. et al., *Nature*, 298, 156–159, 1982.

Note: Pg, petagram, 10[15] g.

a Plant C is one-half the amount of phytomass as reported for biomes in Table 8.2.

b Organic C to 1 m, or above permafrost.

appreciable amounts of the long-wave radiation. Consequently, these gases (CO_2 and CH_4) trap radiant energy in the atmosphere, much as it is trapped beneath the glass of a greenhouse to cause warming in the greenhouse. Because CO_2 is much more abundant than the other greenhouse gases, it is a dominant factor in regulating Earth's temperature, even though it is much less efficient in absorbing long-wave radiation than some of the other atmospheric gases. The global warming effects of atmospheric CO_2 were estimated more than a century ago, but not widely recognized until C.D. Keeling began monitoring atmospheric CO_2 in 1958 on Mauna Loa, Hawaii (Sundquist and Keeling 2009). Subsequently it has been learned that atmospheric CO_2, which had fluctuated between 200 and 280 g/Mg (ppm) for centuries, has increased from 280 ppm only 200 years ago and 320 ppm about 50 years ago to nearly 400 ppm today. The rate of increase is accelerating (Figure 11.6). Initial carbon losses were mainly from deforestation and increased cultivation of land, which reduced the amounts of organic carbon in aboveground biomass and in soils, and more recently losses from the combustion of fossil fuels (coal, petroleum, and natural gas). Cement production is another major source of CO_2.

Combustion of fossil fuels and land use changes, including deforestation, have been adding about 1% to the amount of CO_2 in the

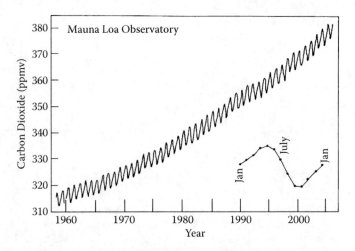

Figure 11.6   Atmospheric CO_2 measured on Mauna Loa, Hawaii. The annual trend is shown on the lower right. The concentrations of CO_2 rise each year until plant growth in the northern hemisphere reverses the trend in May.

atmosphere each year, but the observed increases in atmospheric CO_2 have been considerably less. The main sinks of CO_2 that have taken up much of the excess CO_2 have been the oceans, and possibly reforestation in the northern hemisphere. The oceans have great capacities for accumulating carbon as bicarbonate ions in seawater and as carbonates in the skeletons of microorganisms and small animals in plankton. Warming of the oceans will reduce the capacities of oceans to retain bicarbonate ions in the few thousand meters above the thermocline, and mixing of water into the cold abyss below is extremely slow. Much of the biological productivity in the oceans is on continental shelves and other shallow marine platforms where the water is less than about 200 m deep and is mixed by ocean currents and waves. Upwelling from deeper currents such as the Humboldt and Benguela Currents (Chapter 5) brings nutrients up from below and enhances productivity near the deep currents. Warming of the oceans and melting of polar ice (ice water is not as dense as salty seawater) will reduce the sinking of polar surface water, which is a major means for mixing ocean water and initiating deep currents. Warming of the oceans above the thermocline might increase the amounts of skeletal carbonates retained in bottom muds, because carbonates are less soluble in warmer water, unless this trend is offset by the limitations of increased ocean acidity that reduces the production of skeletal carbonates and makes them less resistant to dissolution.

Global warming has not been obvious because increases in temperatures have been more erratic than persistent; a warm year might be followed by a cold year. We now have records over enough years to show warming trends that appear to be related to anthropogenic releases of excess CO_2, rather than to more natural causes. If one lacks confidence in the weather records, the loss of polar ice in the Arctic Ocean is more graphic evidence of global warming. Greater warming toward the poles than in tropical zones is one of the more definite patterns of global warming. Although we expect changing patterns of precipitation, it is difficult to predict which areas will become wetter and which will become drier, or wetter. The rise of sea level caused by the expansion of seawater as it warms and by water added from the melting of polar ice and glaciers is becoming evident and will result in the flooding of low coastal areas.

An alarming feature of global warming is that it is likely to continue for decades, or possibly for centuries, even if anthropogenic emissions are reduced to preindustrial levels by complete conversion to renewable energy sources. Losses of glaciers and sea ice will decrease the albedo and cause more solar radiation to be absorbed on Earth. Warming of arctic soils and melting of permafrost will allow the decay of the large amounts of organic matter in the soils and the release of great quantities of CO_2 and CH_4 into the atmosphere, frozen soils (Gelisols) contain about 610 Gt C in the upper 2 m, based on extrapolation from data of Bockheim and Hinkel (2007), although only 393 Gt C in the upper meter of Gelisols was considered in the 1500 Gt C for humus in Figure 11.5. A warmer atmosphere will hold more water vapor, which is an effective greenhouse gas. Another cause for concern is that the loss of polar sea ice and warming of the water may allow the release of methane clathrates, which are hydrated forms of methane with vast stores in the cold polar oceans. The outlook is alarming, because global warming can have serious consequences for life on Earth.

Increased reforestation and changes in land use practices, such as reduced tillage and more efficient crop production, may have appreciable effects on the sequestration of carbon (Leifield 2006). Kern and Johnson (1993) estimated that increasing the amount of cropland from three-quarters with conventional tillage to three-quarters with conservation tillage practices in the United States would add carbon to the soils and save carbon losses by reducing fuel requirements for cultivation that would be equivalent to about 1% of the fossil fuel use from all other sources over 30 years. The effects of reforestation, reduced tillage, and more efficient crop production are likely to be limited to a few decades until gains are balanced by use of the forest, rangeland, and agricultural products. Lal (2009) argues that although conservation tillage and other agricultural practices are not expected to be the only answers to global warming, they are cost-effective and have many benefits, for example, reductions in fuel use for cultivation and reduced commercial fertilizer requirements (the fixation of nitrogen for commercial fertilizer, for example, requires considerable energy).

If increased atmospheric CO_2 increases plant productivity, that will have influences into the future. Some plants may grow better with more CO_2 and some may not grow better; the overall consequences are unknown. It is known more definitely that increased CO_2 will increase water use efficiencies by plants. Consequently, plant growth may be increased in drier areas, causing CO_2 to be added to ecosystems and crops in those areas. Burial of eroded sediments, which may amount to 0.8 or 0.9 Gt of C from soils annually (Schlesinger et al. 2000) (Figure 11.5), is another mode of C sequestration, but that may be largely offset by productivity losses from eroded land; the issue is complex (Van Oost et al. 2009).

Unless anthropogenic sources of carbon are drastically reduced, it may be necessary to sequester substantial carbon by the injection of CO_2 gas or liquid into geologic strata and oceans (Sundquist et al. 2009). Although these engineering solutions may be very expensive, the consequences of elevated atmospheric CO_2 are likely to be much more costly.

11.4 Ancient Climates, Plants, Weathering, and Atmospheric CO_2 and Methane

Climates during the Proterozoic eon ranged widely from very hot to very cold ice ages. The extremes were greater than any have been since vascular plants evolved from about 425 Ma in the Silurian. From the Silurian, the more extreme maxima of atmospheric CO_2 concentrations declined on through the appearance of gymnosperms about 360 Ma during the Devonian, and then angiosperms following the Permo-Triassic mass extinction of 245 Ma. The first soils containing organic matter from plants are Silurian, and the first organic soils containing woody plant fossils date from the Devonian period (Retallack 2001). Modulation of global climatic swings has been attributed to fixation of carbon from CO_2 by vascular plants and weathering of silicate minerals to produce bicarbonate or carbonate ions that are precipitated in forming limestone and dolomite (Berner et al. 1983; Royer et al. 2004). The positive relationship between atmospheric CO_2 and temperatures during the Paleozoic has been questioned, but confirmed by Came et al. (2007).

The concept of weathering as a sink for CO_2 is based on the reactions of organic and carbonic acids with the silicate minerals in rocks and soils. For example, CO_2, from the atmosphere, or from the decomposition of humus, reacts with the silicates to produce Ca and Mg carbonates. A simple example is the reaction of CO_2 with olivine:

$$MgSiO_3 + CO_2 = MgCO_3 + SiO_2$$

But a more realistic example is the reaction of CO_2 with water to produce bicarbonate ions that then react with silicate minerals, for example:

$$CO_2 + H_2O = H^+ + HCO_3^-$$

$$CaAl_2Si_2O_8 + H_2O + H^+ + HCO_3^- = Al_2Si_2O_5(OH)_4 + CaCO_3$$

which represent the weathering of Ca-feldspar to produce kaolinite and calcite. Calcium and bicarbonates (Ca^{2+}, $2HCO_3^-$) are leached from acidic soils, and carbonates are precipitated in alkaline subsoils or in sedimentary basins or are washed to oceans where Ca-carbonates are used by marine organisms in the construction of their skeletal structures. Calcite or aragonite from the skeletons of organisms that sink to the bottoms of shallow seas (shallow, because $CaCO_3$ is soluble below about 1500 m) accumulate to form limestone. Although weathering and sequestration of carbon by precipitation of calcium carbonate is very effective in reducing excess levels of carbon dioxide from the atmosphere, it is a very slow process.

A very warm period about 58 Ma, in the Paleogene, has been attributed to large releases of methane from unknown sources (Dickens et al. 1997). Elevated levels of CO_2 may have been important also in causing the high Paleogene temperatures, considerably higher than current temperatures. Temperatures declined through the later part of the Paleogene, and an Antarctic ice cap developed during the Oligocene. Following higher temperatures in the middle of the Miocene, a few degrees Celsius higher than those today, temperatures declined through the remainder of the Neogene. By the end of the Pliocene, Earth had cooled sufficiently that climatic cycles related to the orientation and position of Earth relative to the sun became evident. The major cycles are a 100 ka cycle of eccentricity in Earth's orbit around the sun, a 41 ka cycle of tilt of Earth's rotational axis (22.1 to 24.5° tilt, now at about 23.4°) from perpendicular to Earth's orbit around the sun, and a precession of the equinoxes on a 26 ka cycle that, combined with precession of the elliptical orbit, yields a 21 ± 2 ka cycle.

Analyses of air from bubbles in a 3310 m ice core from Vostok Station in Antarctica provide a detailed record of atmospheric gas concentrations over the past 423,000 years (Petit et al. 1999). Air temperatures and global ice volumes have been inferred from deuterium and ^{18}O concentrations in the gases. The data show similar patterns for each of the last four glacial cycles, with abrupt terminations of glacial climates to the warmer temperatures of interglacial periods every 100 ka. Concentrations of CO_2 and CH_4 are very closely related to the air temperatures (Parrenin et al. 2013). Because rising air temperatures were not preceded by higher CO_2 or CH_4 concentrations, the concentrations of these gases were not causes of the glacial temperature cycles. The atmospheric CO_2 concentration ranged between about 190 ppm during cold stages and 280 ppm during interglacial periods. The rise of CO_2 far above 280 ppm is a recent phenomenon, without precedent during the Pleistocene. Currently rising global temperatures are caused by rising concentrations of CO_2 and other greenhouse gases. Another greenhouse gas that will contribute greatly to global warming is N_2O (Canfield et al. 2010).

11.5 Nitrogen Cycle

Nitrogen in soils and its role in plant metabolism and nutrition are discussed in Chapter 8. The atmosphere is the largest reservoir of nitrogen. There is twice as much N in the atmosphere as in the continental crust (Galloway 2004) and about a million times more than in all forms of N in the living organisms on Earth.

Nitrogen in the atmosphere is in the inert dinitrogen (N_2) form. Small amounts of N are oxidized to N_2O and NO by lightning, and automobiles and industrial pollution are sources of N oxides (NO_x). These oxides and nitric acid produced by hydrolysis of N oxides are washed out of the atmosphere by precipitation or collect on the ground as dry deposition. Most of the N_2 that is converted to more reactive forms is "fixed" by microorganisms near the ground surface, in water, or in soils. Plants use the N fixed in available forms by microorganisms. Detritus from the plants is the source of litter on soils and soil organic matter that contains considerable N.

Plants use about 1.2 Gt of nitrogen each year. About 12% of that nitrogen is supplied by fixation from N_2 by microorganisms, and

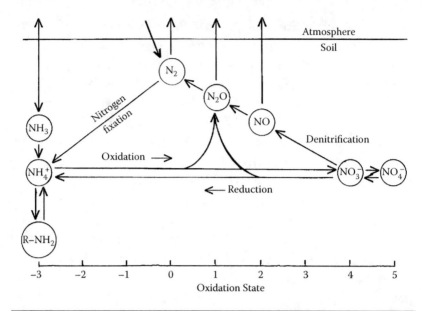

Figure 11.7 Soil nitrogen transformations. These transformations entail processes that are active in both terrestrial and aquatic environments. Some plants take NH_4^+ from soils and some require NO_3^-. (Modified from Figure 1 of Galloway, J.N., in W.H. Schlesinger (ed.), *Biogeochemistry, Treatise on Geochemistry*, Vol. 8, Elsevier-Pergamon, Oxford, 2004, pp. 557–583.)

microorganisms oxidize much of it to nitrate (NO_3^-). The remainder of the ammonium and nitrate assimilated by plants is supplied by the recycling of nitrogen within plant-soil, or plant-substrate, environments. Some nitrogen is returned to the atmosphere by microbial denitrification and by wildfires. A relatively small amount of N_2O that is produced in the denitrification process enters the atmosphere. Although inert in the troposphere, some N_2O escapes to the stratosphere where N_2O oxidized to NO mediates the destruction of ozone. Even though NO is important in soils only as a transient, it is shown in Figure 11.7 for a more complete N cycle.

Fertilizers are substantial components of the nitrogen cycle, and industrial and motor vehicle emissions are very important in large urban areas and downwind from them. More N is added to ocean and terrestrial systems from anthropogenic sources (160 Tg N/year) than is fixed by living organisms in either ocean (140 Tg N/year) or terrestrial (110 Tg N/year) environments alone (Gruber and Galloway 2008). Excessive nitrogen fertilization and nitrates from feed lot effluent have been known to contaminate groundwater and make it unsuitable for drinking. Losses of nitrogen in runoff water are very low in undisturbed ecosystems. Hubbard Brook watershed studies (Likens et al. 1977), however, have shown marked increases in stream nitrogen concentrations within a few years following forest clearing and burning. As the forest became reestablished and recycled more of the nitrogen released by decomposition of plant detritus, stream nitrogen concentrations gradually returned to very low levels.

Most of the nitrogen cycling in natural environments is among plants, microorganisms, and soils (or water) within ecosystems and between the atmosphere and ocean and terrestrial ecosystems, with relatively minor stream transfers on the order of 15–30 Tg N/year from continents to oceans (Jaffe 1992). Excessive fertilization in agricultural areas can increase the runoff of N and P in streams sufficiently that upon reaching the sea, they greatly enhance phytoplankton production. Oxidation of phytoplankton detritus that sinks in shallow coastal water depletes the oxygen, producing "dead zones" where there is too little oxygen for fish, crustaceans, and other animals to survive.

11.6 Sulfur Cycle

Sulfur is an essential element for plants and animals (Bowen 1979). The element occurs in several oxidation states from −2 to +6, and its reactions and compounds are highly diverse.

Most of the sulfur in Earth's crust is in rocks (Table 11.4), primarily as sulfides (for example, pyrite), and secondarily as gypsum. Much of the sulfur that has been released by weathering of igneous rocks and through volcanic activity has been cycled back to the lithosphere in sediments and sedimentary rocks. Most of the S that has not been cycled back to the lithosphere is in the oceans (Table 11.2). Following the rise in atmospheric oxygen about 2 billion years ago, sulfate (SO_4^{2-}) has been the dominant form of S in the oceans. Hydrogen sulfide (H_2S) and dimethyl sulfide (CH_3SCH_3) are commonly released to the atmosphere and rapidly oxidized to sulfur dioxide (SO_2) in dry air and sulfate (SO_4^{2-}) in aerosols and clouds. The sulfate is returned to the ground or oceans in precipitation. Although emissions of carbonyl sulfide (OCS) from oceans, wetlands, and wildfires are much less than those of hydrogen sulfide and dimethyl sulfide, more of it persists in the atmosphere, because it is more stable in oxidizing environments.

Table 11.4   Major Sulfur Reservoirs

Reservoir	Mass of Sulfur (Tg)	Major Forms
Atmosphere		
Continental	1.6	OCS, SO_4^-, SO_2, DMS, H_2S
Marine	3.2	OCS, SO_4^-, SO_2, H_2S, DMS
Biomass and Soil		
Continental	300×10^3	Organic matter
Marine	30	Organic matter
Lakes and rivers	300	SO_4^-
Seawater	1300×10^6	SO_4^-
Ocean sediments	300×10^6	Gypsum, pyrite (FeS_2)
Earth's crust	$24{,}000 \times 10^6$	Sulfides, gypsum ($CaSO_4 \cdot 2H_2O$)

Source: Data from Charlson, R.J. et al., in S.S. Butcher et al. (eds.), *Global Biogeochemical Cycles,* Academic Press, San Diego, 1992, pp. 285–300.
Note: $Tg = 10^{12}$ g.

Oxidation and reduction of sulfur compounds in terrestrial and aquatic environments are the specialities of a variety of bacteria that function in a broad range of environments from very acidic to neutral. Even within the genus *Thiobacillus* different species operate in diverse environments from pH 2 to 7 or 8.

Much of the S in runoff from land to the oceans is derived from residues generated by burning of fossil fuels, which contribute to the dust and wet deposition factor in Figure 11.8.

More S is delivered to the oceans than can be accounted for, although the numbers are balanced in Figure 11.8. It is uncertain whether the S concentration in seawater is increasing or more S is being precipitated as gypsum and sulfides than has been accounted for in this illustration of the S cycle. A large variable is the output from volcanoes, which in some years can be many times greater than shown in Figure 11.8 (Legrand and Delmas 1987).

Most of the sulfur in soils is in organic matter. Plants acquire sulfur as sulfate (SO_4^{2-}). Few soils have appreciable amounts of SO_4^{2-}, because it is readily leached from most of them. Microorganisms decomposing soil organic matter make some SO_4^{2-} available for plants.

Anthropogenic sulfur dioxide (SO_2) emissions, along with nitrogen oxides, are major sources of air pollutants. Coal-burning power plants are particularly large sources of SO_2 that is oxidized and hydrolyzed to produce acidic aerosols and acid rain.

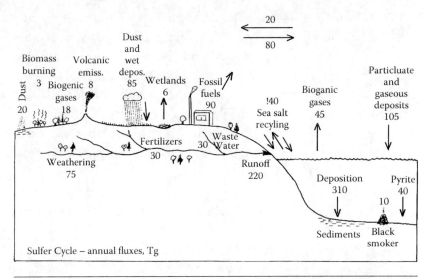

Figure 11.8 Sulfur cycle. The annual global sulfur cycle. Fluxes in Tg/year. (Parts from Schlesinger, W.H., *Biogeochemistry: An Analysis of Global Change*, Academic Press, San Diego, 1997; Berner, K.B., and R.A. Berner, *Global Environment: Water, Air, and Geochemical Cycles*, Prentice Hall, Upper Saddle River, NJ, 1996; and Brimblecombe, P., in W.H. Schlesinger (ed.), *Biogeochemistry, Treatise on Geochemistry*, Vol. 8, Elsevier-Pergamon, Amsterdam, 2004, pp. 645–682.)

11.7 Acid Rain, Water, and Soils

Acid rain is considered to be rain of pH < 5.6, where 5.6 is the pH of pure water in equilibrium with CO_2 in the atmosphere. This pH is down from preindustrial pH 5.7 water. Naturally, rain can be acidic where abundant vegetation emits acid-forming compounds, or it can be nonacidic, where there is enough atmospheric dust to neutralize the acids. The acidity of rainfall unadulterated by atmospheric pollutants has been increasing since the greatly intensified burning of coal and other fossil fuels was initiated during the industrial revolution in the nineteenth century (Howells 1995). Although the pH scale was not developed until the twentieth century (Sorensen 1909), acidity could be evaluated previously by titration of acidic water with dilute sodium hydroxide to a colorimetric endpoint.

Natural sources of rain and water acidity are gases and fluids from volcanic activity, water draining from areas with sulfide deposits, wildfires, and oxidation of plant detritus and soil organic matter. Major anthropogenic sources of acid rain and water are burning of coal and other fossil fuels, motor vehicle emissions of nitrogen oxides, cement production, and N fertilizer (Rice and Herman 2012).

Areas downwind from industrial areas are those most severely affected by acid rain, mainly due to sulfur and nitrogen oxides emitted from power plants, factories, and motor vehicles. Sulfur and nitrogen oxides react with water to produce sulfuric and nitric acids. About half of the acidifying atmospheric deposits are dry and half are in rainfall. Most ammonia (NH_3) deposition is dry. The reaction of ammonia with water neutralizes acidity, but the oxidation of ammonium (NH_4^+) produced in the reaction creates acidity. Acid rain has been a major problem in Western Europe, eastern North America, and Eastern Asia. The acidity of rain has been decreasing since 1980 in Europe and North America, but it is increasing in China (Berner and Berner 1996).

Prior to 1980, rainwater pH was < 4.2 downwind from the industrial Midwest in the United States. Nevertheless, the main source of stream acidity in tributaries to the upper reaches of the Ohio River was water draining from Allegheny Plateau coal mines, which exposed sedimentary strata containing sulfides. According to Pearson (1992), the upper Ohio River was very strongly acidic, while the lower Ohio River running across sedimentary strata containing more carbonates was neutral to slightly alkaline. Although few large rivers are naturally acidic, black water rivers containing considerable organic matter can be very strongly acid; the Rio Negro in the Amazon River Basin and the Satilla River on the Atlantic Coastal Plain are examples of very strongly acidic black water rivers. The Ogeechee is another river with black water on the Atlantic Coastal Plain, but whereas the Satilla River is entirely on the Coastal Plain, the Ogeechee River has headwaters in the Appalachian Mountains (Meyer 1992), where rock weathering and soils supply enough basic cations to the Ogeechee River such that it is only slightly acidic downstream where it is a black water river on the Coastal Plain.

Cation exchange in soils and weathering are major modifiers of water from acid deposition (Dahlgren 2007). Deep, well-drained soils with plenty of clay or organic matter, but not with thick layers of detrital organic matter as in O horizons and histic epipedons, are the most effective in ameliorating the acidic influences of acid rain. Soils with very strongly weathered parent materials and very low cation-exchange capacities lack the cations required to neutralize water that becomes acid as it falls through the vegetative canopy and leaches through the detrital ground cover. Soils, streams, and lakes in areas with granitic rocks are more susceptible to acidification than those in areas with mafic rocks, such as gabbro, because the granitic rocks have less amounts of basic cations and more aluminum. Also, most of the minerals in mafic rocks are more readily weatherable than those in granitic rocks. Terranes with clastic sedimentary rocks have broad ranges of susceptibilities to acidification, depending on the compositions of the sediments. Soils and water bodies in the Adirondack Mountains and central Sierra Nevada of California are quite susceptible to acidification, but those in the Sierra Nevada have received much less acidic deposition (Chen et al. 1991).

The largest area in North America with streams and lakes severely affected by acid deposition is in the Adirondack Mountains. Bedrock in the Adirondack Mountains is metamorphosed Proterozoic granitic, gabbro, and sedimentary rocks, with gneiss from granitic rocks being the most extensive. Since 1979, the sulfate concentrations in wet deposition have declined, but there have been no significant changes in the mean nitrate concentrations. The mean pH of water in the Adirondack lakes has increased about 5.0 to about pH 5.7 (Driscoll et al. 2003). Basic cations have been leached from the soils along with sulfates and nitrates, but extra basic cations from the soils and a shift from inorganic to organic aluminum in the water have increased the anion neutralizing capacity in the water in many lakes, especially in watersheds with thin glacial drift, rather than thick glacial deposits (Driscoll et al. 2003). The inorganic aluminum has contributed to the decimation of fish populations in Adirondack Mountain lakes.

The deleterious effects of atmospheric sulfur and nitrogen oxide acid deposition on plants are generally greater than the effects on animals, although animals are highly sensitive to ozone pollution (Wellburn 1994). A common visible symptom of acidic deposition on plants is bleaching by reactions with pigments in their leaves, causing them to turn white. Aluminum becomes soluble in aqueous environments below about pH 4.8 to 4.5 and is very harmful to fish (Wellburn 1994).

11.8 Phosphorus Cycle

Concentrations of phosphorus are very low in the lithosphere and atmosphere, yet P is one of the most important elements for living organisms.

The primary source of phosphorus is apatite, which has a mean abundance of about 1 g/kg in the lithosphere, primarily in igneous rocks. The concentration of P in the upper continental crust is about 0.6 g/kg (Li 2000). The chemical formula of apatite is $Ca_5(PO_4)_3OH$, with F and Cl substituting for OH in various proportions. Hydroxyapatite forms in bones and fluorapatite is a component of teeth. Biogenic apatite that accumulates in phosphorites of sedimentary rocks is the main source of phosphorus for P fertilizers; guano (excrement from seafowl) is a minor source of P used locally (Jahnke 1992).

The phosphorus cycle is quite different from the nitrogen and sulfur cycles in that phosphorus is present in only one oxidation state, and it forms no gases that are stable in the biosphere or atmosphere. Also, in contrast to nitrogen and sulfur, substantial proportions of the phosphorus in soils are in inorganic forms.

Weathering of apatite releases about 10 Mt of phosphorus annually. In soils monobasic ($H_2PO_4^-$) and dibasic (HPO_4^{2-}) phosphates are generally available to plants. Phosphates are precipitated by calcium in alkaline soils, and most of the phosphate is adsorbed on aluminum and iron oxides in acidic soils. Phosphates are most readily available in slightly acidic to neutral soils. Much of the phosphorus in surface soils is in organic matter. This phosphorus is used over and over again by recycling it through plants and organisms that decompose the plant detritus. Little phosphorus is lost by leaching through soils, but erosion losses on soil particles and plant detritus carried off to aquatic systems may be substantial.

Phosphorus availability is a major factor limiting biomass production in both terrestrial and aquatic ecosystems. Mycorrhiza are efficient scavengers of phosphorus for plants growing in soils with limited availability of the element. Phosphorus fertilization on agricultural land can have detrimental effects when increased phosphate in runoff accumulates in aquatic systems. It can cause increases in aquatic plant and algal growth. If decomposers of the plants and algae use practically all of the oxygen in the water, it becomes an unsuitable habitat of fish and other aquatic animals. The process of abundant nutrient-induced biomass production in lakes and rivers and its decay to deplete the water of oxygen is called *eutrophication*.

11.9 Nutrient Cycling and Plant (Site) Productivity

All vascular plants require 16 chemical elements, including H, O, and C, and at least 5 more are required by some plants. The concentrations of these elements, plus Al and F, in continental crust, soils, and plants are shown in Figure 11.9. Animals require F, along with most of the elements required by plants, plus I and Se. Molybdenum is important because it is required by bacteria in the reduction of nitrates in plants. Plants contain much higher concentrations of C and N than soils, and higher concentrations of S (Figure 11.9). These are the major elements cycled through soils, and the atmosphere is important in these cycles. Rocks and soils are the main sources of the other essential elements, and weathering (Sections 2.2 and 11.1) is important in

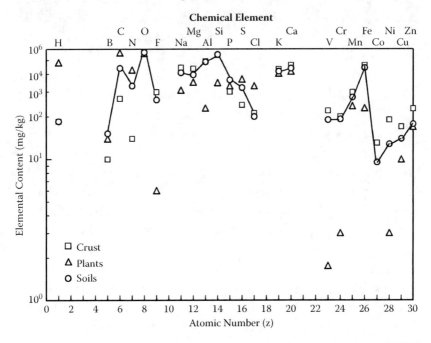

Figure 11.9 Concentrations in the continental crust, soils, and plants of the chemical elements among the first 30 (atomic numbers 1 to 30) that are most important for plants and animals. The mg/kg units are equivalent to parts per million (ppm) by mass: 10^x is equivalent to 1 followed by x zeros. (From Romankevich, E.A., in V. Ittekkat et al. (eds.), *Facets of Modern Biogeochemistry*, Springer-Verlag, Berlin, 1990, pp. 39–51, for continental crust and plants, and Shacklette, H.T., and J.G. Boerngen, *Element Concentrations in Soils and Other Surficial Materials of the Conterminous United States*, Professional Paper 1270, U.S. Geological Survey, Reston, VA, 1984, for soils.)

releasing these elements from rock and soil minerals into soils where they can become available to microorganisms and plants.

Both soils and plants are very important in the cycling of not only S and P, but also basic (alkali and alkaline) cations with mean concentrations > 10 mmol/kg in plants (Figure 11.9). These basic cations are Na, Mg, K, and Ca. They are soluble in soil solutions, except that Ca and Mg are precipitated in alkaline solutions. Otherwise, the basic cations must be incorporated into clay minerals or organic matter, or retained on the cation-exchange complexes, or they will be leached from any soils that have water draining from them. Magnesium is a major constituent and K a minor constituent of some clay minerals, but the main nonleachable retention of the basic cations is in organic matter. Plants contain much more Ca and K than Mg and Na, ensuring that more of these elements will be retained in soil-plant systems. More of the Mg and Na are leached from soils and carried to interior basins or oceans. Consequently, there is much more Na and Mg than K and Ca in oceans. Much of the Ca and considerable Mg that reaches oceans are precipitated as carbonates (calcite in limestone and both Ca and Mg in dolomite), and in interior basins some of the Ca is precipitated as a sulfate (gypsum). Sodium and K salt deposits are much less common.

Silicon and Cl are either so highly concentrated in soils and are available to plants, or of such minor importance to plants, that they are never considered to be elements that limit plant growth or survival. Iron and Mn, which are the other chemical elements with concentrations > 1 in soils and plants, seldom limit plant growth. However, Fe is commonly a growth-limiting element for phytoplankton in marine ecosystems. These elements (Fe and Mn) are mobile as divalent cations, but not in higher oxidation states. They are retained in soils mainly as oxides and oxyhydroxides and are only sparingly available to plants, but plants do not require much of them. Iron oxides accumulate in most soils as they age and impart yellowish brown (goethite) and reddish (hematite) hues to them.

Questions

1. Some continental crust is billions of years old, but no oceanic plates are more than 210 million years old. Why are no ocean plates older than 210 million years (210 Ma) old?

2. How is water added to soils?

3. How is water lost from soils?

4. Carbon dioxide (CO_2) and methane (CH_4) are greenhouse gases released from the decomposition of organic matter. Which is the dominant gas from the decomposition of organic matter in well-drained soils? Which is commonly released from wet, anaerobic soils? Which is the more potent greenhouse gas, absorbing more short-wave energy per molecule? Which is causing the most global warming, because it is much more abundant?

5. Why is global warming in polar regions more alarming than global warming in tropical regions?

6. According to the data in Table 11.2, there is 72 times more C than N in the oceans. Does this indicate that more of the carbon is in organic matter or in carbonates? The numbers in Table 11.2 do not include the vast store of C in ocean sediments.

7. What land use practices will have the greatest effects in promoting the accumulation or sequestration of organic carbon? Which will reduce the use of fossil fuels?

8. How might soil erosion affect the global C balance?

9. Where do plants colonizing a recent glacial deposit or a deep deposit of volcanic ejecta (tephra) obtain N—from the incipient soil parent material (glacial till or volcanic ejecta) or from the atmosphere?

10. Can the colonising plants alone obtain the N that they need, or are they dependent on bacteria, such as cyanobacteria or the bacteria in lichens, to acquire N to initiate colonization?

11. In stream waters of the Adirondack Mountains, downwind from the industrial Midwest, S is decreasing, but N is not. Why is the N in acid rain over the Adirondack Mountains not declining as the emissions from power plants become cleaner?

12. Why are soils with gabbro and basalt parent materials more effective in ameliorating the effects of acid rainwater than are soils with granitic parent materials?

13. If the P ions and complexes are relatively immobile in soils, why are fungal mycelia so important in scavenging P for plant use in soils with low P concentrations?

14. For what plant nutrient elements other than C, H, and O is Earth's atmosphere an important link in global cycling?

Supplemental Reading

Berner, K.B., and R.A. Berner. 1996. *Global Environment: Water, Air, and Geochemical Cycles*. Prentice Hall, Upper Saddle River, NJ.

Condie, K.C. 2011. *Earth as an Evolving System*. Academic Press, Boston.

Coyne, M.S. 1999. *Soil Microbiology: An Exploratory Approach*. Delmar Publishers, Albany, NY.

Hoffman, K.A. 1988. Ancient magmatic reversals: clues to the geodynamics. *American Scientist* 256: 96–103.

Li, Y.-H. 2000. *A Compendium of Geochemistry, from Solar Nebula to the Human Brain*. Princeton University Press, Princeton, NJ.

Romankevich, E.A. 1990. Biogeochemical problems of living matter of the present day biosphere. In V. Ittekket, S. Kempe, W. Michaelis, and A. Spitzy (eds.), *Facets of Modern Biogeochemistry*, pp. 39–51. Springer-Verlag, Berlin.

Schlesinger, W.H. 1997. *Biogeochemistry: An Analysis of Global Change*. Academic Press, San Diego.

Smil, V. 2002. *The Earth's Biosphere*. MIT Press, Cambridge, MA.

Taylor, S.A. (1982). *Solar System Evolution*. Cambridge University Press, Cambridge.

12

LAND MANAGEMENT
AND SOIL QUALITY

Land management entails managing the plants and animals that occupy the land. We are dependent on those plants and animals, and they are dependent on soils. The plants are rooted in soils, and the animals depend on plants for their sustenance. Also, soils are regulators of water flow and water quality.

Ecological principles can be applied at any level, or intensity, of management. They are more commonly applied on land that is less intensively managed, including large tracts of forest and rangelands. These forest and rangelands are managed for timber and other forest products, domesticated grazing and browsing animals, and wildlife. Recreation may be a primary or a secondary benefit, and ecological land management helps to maintain plant and animal diversity. Although soils are a more important issue in forest and rangeland management, soil erosion, and compaction, soils are an important consideration in recreation areas also.

12.1 Types of Use and Management for Different Lands

We are dependent on land for dwelling space, food, building materials, and fiber for clothing and other purposes. As the population on Earth increases, it becomes increasingly important to match land to its most appropriate uses. Many patterns of land management are fixed by current urbanization and mining operations. Some choices of land use can be made in most areas, although the choices might be quite limited in densely populated and intensively cultivated areas.

It is prudent to concentrate intensive agriculture on easily cultivated land with readily available and maintainable plant nutrient supplies and with minimal compaction and erosion tendencies. Climate is an important consideration also. For example, land in arid climates is suitable for cultivation only if water is available for irrigation, and water can be a scarce commodity.

All very steep land (slopes > 65%) is unstable when the vegetative cover is drastically disturbed and should not be managed intensively. A slope of about 62 to 70% (32 to 35°) is a critical threshold, because that is near the angle of repose for most soil materials. Very steep land can be managed for the scattered harvesting of individual trees by helicopter or as range for grazing or browsing animals, but it is not suited to the use of mechanized equipment. Grazing should not be intense enough to greatly reduce or disturb the ground cover.

Steep land (30–65% slopes) is most suitable for management that leaves a ground cover of plant detritus or herbaceous plants that minimized the erosion hazard. Ground disturbance can reduce the infiltration rate of precipitation, and it exposes soil to erosion and can concentrate water to produce gully erosion. Cultivation for orchards may be a suitable practice if a cover of grass or coarse detritus is maintained between trees, shrubs, or vines. In some intensively populated areas, terraces have been constructed for cultivation.

Some moderately steep land (15 to 30% slopes) is suitable for cultivation that is directed across slopes, along contours, rather than up- and downslopes. Slope stability, especially erodibility, should be taken into account in choosing the type and intensity of cultivation.

Slightly to moderately or steeply sloping land (3–15% slopes) can be managed either as recommended for moderately steep land, or similar to the management of level land, depending greatly on the erodibility of the soils and the erosivity of the rainfall.

Level, or nearly level, land offers the most management flexibility, if there are no drainage or serious soil structure problems or severe compatibility. In areas of favorable climate, much of the land that has supported forests of broadleaf trees or mid-latitude grasslands has been cleared for cultivation. Savannas, marshes, and shrublands that lack forests, because of shallow soils, dry climate, or soils lacking adequate plant nutrients, are generally more suited to grazing by domesticated animals or for wildlife management. Large areas of cultivated land should include refuges for wild animals and wildlife and corridors connecting the wildlife refuges.

Riparian areas should be managed for the aquatic habitat, and many riparian areas make suitable wildlife corridors. Some floodplains may be cultivated for annual plants or managed for the harvest of perennial wetland plants. Urban development on active floodplains is impractical, except for port facilities of navigable streams. It requires flood control by building dams, diverting watercourses, or building levees.

12.2 Water

Water is essential for the existence of life. The amounts and distributional patterns of water are extremely important in determining the kinds and quantities of biotic communities occupying the land. Management of the vegetative cover can alter the amount and quality of water that reaches the ground surface, near surface air currents and atmospheric humidity, and evapotranspiration. Land management can affect the amount of water that enters a soil (infiltration), the flow of water in the soil, and the amount of water retained in soils. Examples of extreme effects of land management, which includes management of the vegetation, are hill slope failure and surface erosion. Natural or management-induced mass failures on very steep slopes are commonly debris avalanches (Figure 12.1), and the mass failures on moderately steep slopes with fine sediments or soils are commonly slumps and earth flow (Figure 12.2). Surface erosion by overland flow of water is generally called *sheet erosion* if it is superficial or from shallow ephemeral channels called rills (Figure 12.3), or it is called *gully erosion* if the run-off channels are deep and permanent (Figure 12.4). In drier climates, *desertification* and *salinization* are extreme effects that are commonly exacerbated by poor land management.

Water reaching a soil or its vegetative cover is retained on the plants (interception), enters the soil (infiltration), or runs off over the land surface (Figure 5.1). Water that infiltrates into a soil is either stored in the soil or drains through to a stream or to the groundwater table. Water that remains in a soil after runoff overland and drainage through the soil restores the soil water content to the soil's field capacity.

The amount of rainfall (or melted snow) retained by interception on vegetation ranges from less than 0.02 cm for grassland and chaparral to about 0.15 cm in some forests (Kittredge 1948; Satterlund and Adams 1992). This interception is no more than 2.5% of the precipitation from a 6 cm rainfall event, but up to 15% from a 1 cm rainfall event. Most of the intercepted precipitation is evaporated and returned

Figure 12.1   Mass failure in very steep to steep terrain. A debris avalanche commenced on a very steep slope, many years after the forest had been clear-cut. (From Sidle, R.C., *Slope Stability on Forest Land*, Pacific Northwest Extension Publication PNW 209, Oregon State University Extension Service, Corvallis, 1980. With permission from the Oregon University Extension Service.)

to the atmosphere. Precipitation that falls through a canopy of vegetation to reach the ground below is commonly called throughflow.

Practically all of the precipitation that flows through well-vegetated plant communities to the ground infiltrates into the soil, except in extreme rainfall events on saturated soil—exceptional conditions that do not occur every year, or even every decade, in most areas. In contrast, much of the precipitation on compacted or crusted soils lacking full vegetative cover runs off overland rather than being infiltrated into soils. Algal and lichen crusts provide some soil protection from raindrop impact and increase infiltration, reducing overland runoff and erosion.

12.3 Soil Degradation

Three major cause of soil deterioration, or diminishing soil quality, are loss of organic matter, compaction, and erosion. Also, salinization

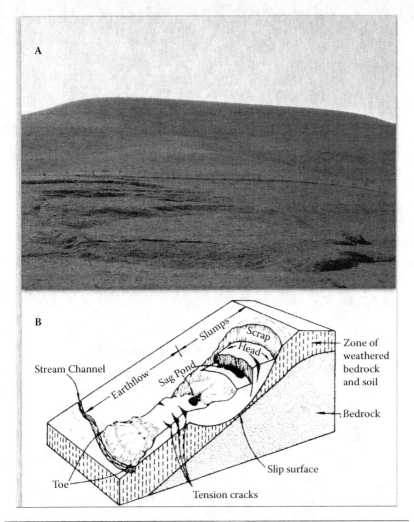

Figure 12.2   Earth slumps. (A) An active slump in intensively grazed grassland of the California Coast Ranges. (Photo by author.) (B) A diagram of a slump shown as a rotational mass failure, with earth flow downslope from the slump. (From Sidle, R.C., *Slope Stability on Forest Land*, Pacific Northwest Extension Publication PNW 209, Oregon State University Extension Service, Corvallis, 1980. With permission from the Oregon University Extension Service.)

is a major cause of soil deterioration in arid areas where insufficient water passes through soils to leach excess alkali cations (mainly Na^+) from them. Loss of soil organic matter (SOM) is an inevitable result of management that involves the conversion of natural vegetation to commercial plants that produce less biomass, or biomass that is harvested periodically. Soil compaction and erosion, however, can be avoided, or at least minimized, by good management practices.

Figure 12.3   Sheet and rill erosion by overland flow of storm water on farmland. (From Soil Survey Staff, *Soil Taxonomy: A Basic System for Making and Interpreting Soil Surveys*, Agriculture Handbook 436, USDA, Washington, DC, 1999.)

12.3.1 Loss of Organic Matter

Although much plant biomass is commonly removed in forest and crop harvest, large amounts of plant biomass are generally left on the ground. Allowing the unharvested biomass to accumulate in surface organic layers, and the products of its decay to return to the soil, minimizes soil organic matter losses.

12.3.2 Compaction

All soils can be compacted, that is, reduced in volume. Soils are most vulnerable to compaction when they are moist, near field capacity, and without a cover of vegetation or plant detritus. It is the macropores that are filled with water when soil is saturated and vacated when a soil is drained at field capacity that are compressed most severely. These are the pores that, upon draining, allow air to circulate through them. The changes in soil porosity are generally detrimental to plants, but the changes can have beneficial effects in very porous soils where drought is more of a threat than restricted drainage.

Mechanical resistance to the stresses of soil compaction is called *soil strength*. Dry soils are the strongest, and their strengths decrease with increased water content. Clayey soils that are very strong when dry may not be any stronger than other soils when wet. Wet soils that are completely saturated cannot be compacted, because water is

Figure 12.4   Gully erosion of forestland following the denudation of ponderosa pine forest by fumes from a copper smelter. Some pine stumps remain between gullies and whiteleaf manzanita (*Arctostaphylos viscida*) is reoccupying the soils. (From Wieslander Vegetation Type Mapping Collection, Marian Koshland Bioscience and Natural Resources Library, University of California, Berkeley.)

practically incompressible. Wet soils can be deformed, however, and compressive and shear stresses on wet clayey soils can cause *puddling*, which is loss of structure. Management operations on wet soils can be direct causes of puddling, or the exposure of bare soil to raindrops can result in puddling. Because of strong cohesive forces between clay particles, spaces between the soil particles are reduced upon drying of puddled soils. Consequently, puddled soils are poor conduits for water and air. It may take years for puddled soils to regain their previous structures. Reduced infiltration rates through puddled soils may lead to more runoff of water, and increased overland flow is likely to increase soil erosion losses.

Soil compaction is sometimes related to the pressures that machinery and animals exert on soils. This is somewhat misleading; for example, the pressure (force/area) of a bicycle tire is several times greater than that of a rubber-tired tractor, yet the compaction caused by a tractor is much greater. Obviously, reducing the tire pressure will reduce the pressure, or stress, at each point of contact on the ground surface, but the stress from tractor tires will be applied over larger areas. Because a tractor is much heavier than a bicycle, it will cause greater stresses to much greater depths below the ground surface. In many cases, the ground disturbance caused by tracked vehicles, especially when the vehicles are turning, is more detrimental than the soil compaction.

Overall soil compaction, or reduction in pore space, is commonly measured by ascertaining particle and bulk densities, where the particle density (Dp) is the soil weight divided by the volume of its particles, and the bulk density (Db) is the same weight divided by the total volume of soil, including pore spaces. The total porosity (P) is the total volume of soil minus the volune of soil particles. It is related to the particle and bulk densities by $P = 1 - Db/Dp$ (Chapter 3). Macroporosity is more arbitrarily defined than is total porosity. It is commonly assumed to be the air-filled porosity when the smaller pores are filled at field capacity. This might appear to be an absolute definition, but it is no more absolute than field capacity, which was defined and discussed in Chapter 5. The permeabilities of soils to water and air are indices of macroporosity that can be measured in the field. Reduction of macroposity is a better indicator of possible detrimental effects on plants than is total porosity.

Increase in soil strength is another indicator of detrimental effects that soil compaction might have on plants. It is easy to record by pushing a rod with a cone-shaped point into the ground and noting the resistance, but it is difficult to evaluate because of the dependence of soil strength on soil water content, or water potential. Perhaps the best strategy for comparing soils is to make penetrometer recordings when the soil is near field capacity, that is, after the soil has drained of excess water, but before it has lost much water from evapotranspiration. All soils have water potentials of about 0.01 MPa when they are near field capacity (Chapter 5).

It may take years for compacted soils to regain their natural porosities. Mechanical operations, such as deep tillage, are sometimes applied on intensively managed land to create large pores that increase drainage and facilitate root penetration. Commonly, the best strategy is to allow nature to take its course—perhaps helping by encouraging the growth of deeply rooted plant species. Shrink and swell of clayey soil, freeze and thaw of cold or very cold soils, and plant root and animal activities in soils are means of soil recovery from compaction.

12.3.3 Erosion

Erosion generally means soil loss by overland flow of water or by wind blowing soil away. In humid areas with good vegetative cover, much of the soil loss may be by removal of the products of weathering in water draining from the soils, which is generally a natural process that is not commonly considered in evaluations of soil erosion. Land management can have large effects on losses from the soil surface in wind, flowing water, or by mass failures, such as mudflows and landslides (Figures 12.1 and 12.2), but management effects on solution losses are generally small, or negligible. There are many ways to monitor soil erosion (references in Supplemental Reading at the end of this chapter). Pedestals of soil capped by pebbles or other objects and accumulations of sediment behind obstacles, such as fences or fallen trees, and sediment in slope depressions downslope from eroded areas are obvious visual indicators of slope erosion. Erosion that removes enough soil to expose subsoils can be considered severe.

Surface erosion rates are very difficult to predict, mainly because most of the erosion occurs in exceptional rain or wind storms that might be expected over long periods but cannot be predicted to happen on any given day or even in any particular year. Erosion commonly happens when a soil is disturbed or exposed, and thus most vulnerable to the deleterious effects of water or wind. Consequently, empirical, or black box, approaches to predicting soil losses are practically as good as mechanistic predictions based on process models. The most widely utilized predictive systems are the revised universal soil loss equation (RUSLE) (Renard et al. 1991, a revision from USLE, Wischmeier and Smith) for water and the wind erosion equation (WEQ) of the USDA for wind.

RUSLE is used to predict rill, or sheet, erosion (A). Each prediction of annual soil loss (A) for a small area with approximately uniform slope is the product of five factors, A = RK(LS)CP, where R is a rainfall erosivity factor, K is a soil erodibility factor, LS is a slope length and gradient (steepness) factor, C is a cover factor, and P is an erosion control practice factor. The factor K depends largely on the detachability and transportability of soil particles. Silty soils are generally given higher K values than clayey or sandy soils, because clay particles are less readily detachable and sand requires a stronger flow of water for transport. Sandy soils may be more susceptible to gully erosion, but RUSLE rates only sheet and rill erosion. The slope length part of the LS factor is commonly difficult to evaluate. It may be the distance across a cleared area, rather than the distance water might travel in a rill down an entire slope. If exposed slopes are very long, water transported in rills will either infiltrate into the soil or be concentrated in gullies before travelling long distances. The C factor is based on the amount of ground cover, which in uncultivated areas is generally rock, small plants, and plant detritus. It is practically zero on densely vegetated land. The P factor is for cultivation practices such as cultivation along slope contours, alternating crops in strips, and terracing. It is generally considered to be 1.0 on uncultivated land. In ecological land management the greatest erosion and largest source of sediments reaching streams are commonly from roads and other construction projects. It is very important to ensure that roads are drained properly and that cut banks are protected by rock, plant cover, or erosion control practices such as terracing. Unstable slopes should be avoided as much as possible in road construction.

A computer model called RUSLE2 was developed under the supervision of George R. Foster of the USDA Agricultural Research Service. It is available from fargo.nserl.perdue.edu and is accessible from the Natural Resources Conservation Service (NRCS) website.

Wind erosion is a function of I, C, K, L, and V, where I is a soil erodibility factor based on the size distribution of discrete soil particles or stable soil aggregates, C is a climatic factor based on wind velocity and soil moisture, K is a surface roughness factor, L is a length of wind flow path index that is based on the length and width of exposed surface between covered areas or windbreaks, and V is a vegetative cover index based on standing vegetation and amount of ground covered by rock and coarse plant detritus. Coarse silt to fine sand grains and aggregates of clay are most susceptible to wind erosion. Larger grains require much greater wind velocities for transport, but relatively large sand grains can be moved by bouncing along (a process called *saltation*), rather than being airborne for long distances. Fine sand grains settle in dunes, while finer particles (dust) may be carried long distances before settling out or being washed from the atmosphere in raindrops. Clay particles may be nuclei for the formation of raindrops from atmospheric moisture. Fine to medium sand transported short to moderate distances commonly accumulates around shrubs, leeward from other obstacles, or in sand dunes. Aeolian clay can accumulate in mounds downwind from dry lake beds and ephemerally wet basins or playas with salty soils (Bowler 1973); they are relatively common in southeastern Australia, where the mounds consisting of sheets of clay are called *parna*.

12.4 Diversity of Plants and Animals

Soil and biotic diversity is highly scale dependent. As the area considered increases, the total diversity increases, but the diversity per unit area may decrease.

Different soils in similar climatic regimes may have plant communities that differ only slightly on similar soils to substantially in landscapes with soils having a diversity of limiting conditions, such as shallowness or restricted drainage. Some soil parent materials, such as limestone and serpentine, have soils that support unique plant communities.

The number of species is a common measure of biotic diversity. Is the number of species in a biotic community related to the soil and ecosystem productivity? The species diversity is certainly greater in wet tropical areas than in cold regions. The species-productivity hypothesis suggests that the diversity is reduced on soils with limestone and serpentine parent materials, but a landscape with great soil diversity is likely to have great species diversity; thus, a landscape that contains limestone or serpentine soils along with other kinds of soils may be expected to have relatively great species diversity.

Human activities can degrade soils and reduce biotic diversity. Ecological land management that maintains the hydrological functions and biological habitability of soils will curtail or minimize any losses of plant and animal diversity.

12.5 Soil Quality

12.5.1 Aspects of Soil Quality

Soil quality is a subjective concept that relates to the usefulness of soil in regulating environmental processes and in producing products for human consumption. Most water from precipitation enters soils. Soils store water to field capacity and slowly release the excess water to groundwater and streams, diminishing the incidences and magnitudes of flash flooding in the streams. Soils are important in maintaining water quality by removing toxic substances as water passes through them. Soils have minor effects on climate that can be important locally; for example, they can temper atmospheric extremes of temperature by storing heat when weather is hot and sunny and releasing heat when it is cold and dark, and extremes of atmospheric dryness are damped by moisture released from soils by evaporation or by transpiration from plants. Soils are essential for efficient plant production—little phytomass is produced on bare bedrock. The capacity of soils to provide these watershed, ecological, and agricultural functions is called *soil quality*.

Each soil has a natural, or inherent, capacity to regulate environmental processes and produce products for human consumption. This capacity may be altered by disturbance or soil management, or mismanagement. The status of a soil, relative to its inherent quality under natural conditions, is referred to as its *soil condition*. It is commonly called *soil health*. Soil *condition*, however, seems more appropriate for soils, reserving the word *health* for the condition of animals and other organisms in soils. Soil quality generally refers to the current soil capacity, which may be either more or less than the inherent capacity, depending on whether its capacity has been increased or decreased by management. The capacity of a soil to produce crops may be increased by the incorporation of manure, for example, or reduced by compaction and reduction of porosity.

Because we are dependent on soils for food production and many watershed and ecological functions, the maintenance of sustainable life on Earth requires that we maintain soil quality.

12.5.2 *Indicators of Soil Quality and Soil Condition*

The capacities of soils to yield useful products and regulate environmental processes are related to soil properties such as organic matter content, soil porosity, nutrients available to plants and other living organisms, and the diversity of living organisms in soils. Some of the yields and regulatory functions of soils are difficult to measure or evaluate. Many proxies have been proposed for rating soil quality (Doran and Jones 1996). No one proxy for evaluating soil quality and condition is a suitable indicator for all types of productivity and all kinds of environmental processes. Some indicators are soil properties such as porosity, some are biological indicators (Bloem et al. 2006), and some are estimates of soil degradation by erosion or compaction. Good site assessment for management purposes requires evaluation by more than one indicator.

Some indicators of soil quality are practically fixed and some are more alterable (Table 12.1). Relatively permanent features, such as soil depth, may be good indicators of inherent or potential soil quality, but they are not useful indicators of soil condition. Soil depth and texture are not affected significantly by management or other temporal factors, other than industrial and mining operations and catastrophic events such as slope failure or very severe erosion. Some soil features, such as porosity, can be altered in minutes by the operation of heavy machinery on moist soils. Readily altered features such as soil porosity, or surrogates such as soil permeability, should be considered in evaluating prevalent qualities or conditions of soils that might be subjected to soil-altering events. Some soil properties, such as surface temperature, are subject to diurnal changes, and one recording may not be a suitable indicator for the temperature aspects of soil quality—only means of many observations are suitable for evaluating these properties. Some features, such as subsoil temperature, change in annual cycles and can be evaluated for specific time periods during a year or as annual means. Most biological features are subject to annual, or more frequent, changes.

Among the factors to consider in choosing soil quality and condition indicators and methods are kinds of ecosystems, including ecosystem geology and other landscape features, type of land use or management, and project duration. Some indicators will be more

Table 12.1   Some Indicators of Soil Quality

Soil Property	Method for Evaluation	Days	Months	Annual	Decades	Centuries
Physical Properties						
Soil depth	Probe					x
Texture	Feel, lab tests					x
Strength						
Dry	Feel, crushing	x				
Wet, plastic limit	Feel, malleability				x	
Structure						
Peds	Visual	x				
Aggregates	Stability in water		x			
Porosity						
Total	Bulk density	x				
Macropores	Permeability	x				
Temperature						
Surface soil	Thermometer	x		Cyclic		
Subsoil	Thermometer		x	Cyclic		
Chemical						
CEC	Laboratory				x	
pH	Instrument or indicator solution		x	Cyclic		
Organic C	Lab or visual		x			
Organic N	Laboratory		x	Cyclic		
Phosphorous	Laboratory		x			
Salinity	Electrical conductivity		x			
Biological–Physical or Biological–Chemical Properties						
Microbial biomass			x	Cyclic		
Nematodes			x	Cyclic		
Arthropods			x	Cyclic		
Mites						
Springtails						
Enchytraeids			x	Cyclic		
Earthworms			x	Cyclic		

Note: Methods of estimating microbial biomass have been developed to supplement or replace direct counts of cells; these methods include chloroform fumigation and extraction of carbon and nitrogen constituents from lysed cells, and substrate-induced respiration.

sensitive than others, and the most sensitive indicators will be different in different situations. Indicators that fluctuate on annual or more frequent cycles will need controls if they are to be used on a seasonal basis; for example, soil temperature in a disturbed area can be compared to that in similar undisturbed areas over the same time period. It is important to choose indicators that management might alter within reasonable time periods For this reason, soil texture and soil depth are not likely to be suitable indicators for soil condition, but thickness of the A horizon may be a useful indicator, because it can change over a relatively short time span.

12.6 Degraded Landscapes and Soils

The degradation of landscapes ranges greatly in severity, from landform remolding by strip mining to nonmechanized harvesting of plants. Efforts required to repair degraded landscapes and restore watershed and ecosystem functions range from mechanized operations with heavy equipment to reclaim some of the more severely degraded landscapes (Toy and Hadley 1987) to simply excluding the entry of human and domesticated animals (Table 12.2). Many aspects of land and ecosystem restoration are discussed in Comín (2010).

Some cases where soils are completely destroyed or permanently altered are strip mining, dredging for gold in stream deposits (Figure 12.5), dredging sand and gravel for construction materials, and the urbanization of wetlands. Laws may require restoration or mitigation, but strip-mined land will never be the same as the original, and it is questionable that "made," or artificial, wetlands will ever have functions comparable to the land that they are supposed to replace.

Soil compaction, erosion, and frequent or periodic losses of vegetative cover by harvesting or burning are major causes of soil deterioration. Some of the results are losses of soil organic matter and nutrient storage capacity, loss of soil structure and decreased rates of drainage and air exchange, and ground surface crusting that reduces infiltration and increases overland runoff of water. Adverse effects on soil biota are a common consequence of soil degradation, and *salinization* occurs in some of the more arid areas (Lal et al. 2004). The expansion of arid soil conditions by overgrazing or other management that reduces the

Table 12.2   Reparation of Degraded Landscapes

Reclamation Requiring Heavy Mechanized Equipment

Reshaping mine spoil landscapes
Removing dams from streams
Eliminating roads

Reclamation Requiring Heavy or Light Mechanized Equipment
Restoring water sources to wetlands that had been drained
Spreading soil that was displaced in the removal of plants

Restoration Requiring Light Mechanized Equipment or None
Removal of undesirable plants
Planting of desirable plant species for ground cover and for rehabilitation of degraded surface soils
Planting of deep-rooted plants to alleviate soil compaction
Planting native plant species to achieve diversity
Removal of building structures from floodplains and restoring of natural stream flow

Restoration requiring only desirable management
Exclusion of human and domesticated animals
Reintroduction of native plant propagules and native animals
Prevention of excessive harvesting of trees from riparian areas
Prevention of obstructions to stream flow

Figure 12.5   Tailings from dredging for gold on the floodplain of a Sierra Nevada stream. (From Wieslander Vegetation Type Mapping Collection, Marian Koshland Bioscience and Natural Resources Library, University of California, Berkeley.)

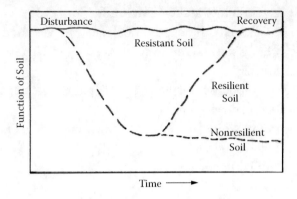

Figure 12.6   Possible soil responses following disturbance.

plant cover and the capacity of soil to support the replacement of that vegetation is sometimes called *desertification* (Geist 2005). Many of the effects of environmental degradation are discussed in Christ and Rinker (2010).

Some soils are more susceptible to degradation than others, and some recover more quickly than others (Figure 12.6). Soil organic matter, cation-exchange capacity, and weatherable minerals are some of the soil features that mediate resistance to soil deterioration and recovery of satisfactory soil condition, or soil quality. Viable biological communities are very important in ameliorating the effects of degradation and hastening the recovery of favorable soil condition, or soil quality. Special erosion control practices (Morgan 2005) may be necessary on bare soils before the plant cover and its detritus (or litter) are adequate to limit erosion. Restoration of the natural functions may take years, or decades, for deep soils, and centuries, or millennia, for shallow soils.

Questions

1. Why is land steeper than 65% (about 33°) unsuitable for cultivation?
2. Why did the debris avalanche of Figure 12.1 occur long (years) after clear-cutting, rather than soon after the trees were cut? Hint: Tree roots add strength to the soils, or regolith.

3. How do erodibility and erosivity differ?

4. Are there rainfall, drainage, and plant cover conditions under which clayey soils are more susceptible to gully erosion than sandy soils?

5. What soil properties increase soil resistance to degradation by compaction and erosion?

6. Which of the following land suitability or land use considerations involve soils?

   a. Proximity to streams

   b. Growth rates of trees

   c. Soil drainage

   d. Connectivity to other areas with natural ecosystems

   e. Length of plant growing season between frost periods

   f. Groundwater quality

7. What soil properties contribute to the resilience of disturbed soils?

8. What landscape reparation activities might be added to Table 12.2?

Supplemental Reading

Bolen, E.G., and W.L. Robinson. 1999. *Wildlife Ecology and Management.* Prentice Hall, Upper Sadle River, NJ.

Christ, E., and H.B. Rinker (eds.). 2010. *Gaia in Turmoil, Climate Change, Biodepletion, and Earth Ethics in an Age of Crisis.* MIT Press, Cambridge MA.

Comín, F.A. (ed.). 2010. *Ecological Restoration: A Global Challenge.* Cambridge University Press, Cambridge.

Doelman, P., and H.J.P. Eijsackers (eds.). 2004. *Vital Soil.* Elsevier, Amsterdam.

Fisher, R.F., and D. Binkley. 2000. *Ecology and Management of Forest Soils.* Wiley, New York.

Heady, H.F., and R.D. Child. 1994. *Rangeland Ecology and Management.* Westview Press, Boulder, CO.

Smith, D.D., and W.H. Wischmeier. 1962. Rainfall erosion. *Advances in Agronomy* 14: 109–148.

Whisenant, S.G. 1999. *Repairing Damaged Wildlands.* Cambridge University Press, Cambridge.

Williams, J.E., C.A. Wood, and M.P. Dombeck (eds.). 1997. *Watershed Restoration: Principles and Practices.* American Fisheries Society, Bethesda, MD.

Glossary

Aerosol: A colloidal suspension of solid particles or liquid droplets in a gas, for example, clouds, dust, and soot in the atmosphere.

Alkalinification: Accumulation of sodium in soils with enough carbonate ions to raise the alkalinity to pH > 8.2.

Alluvium: Terrestrial sediments deposited from running water.

Ammonification: The conversion of nitrogen from organic matter to ammonium (NH_4^+) by microorganisms.

Assimilation: Constructive metabolism—the use of food or nutrients in biological processes within living organisms for growth, reproduction, or maintenance.

Available water capacity (AWC): The capacity of a soil to retain water that is available to plants. It is the amount of water retained between the *field capacity* of a soil and the *wilting point* of plants.

Barren (ecology): An area with scant vegetative cover, or none at all.

Basin-fill: Alluvium deposited in a broad valley by streams flowing along the longer axis of the valley.

Cation: A positively charged ion.

Cation exchange: The interchange of positively charged ions among solids or between solids and solutions surrounding the solids.

Cation-exchange capacity (CEC): The capacity of a soil to retain cations that are replaceable by an alkali or an alkaline earth cation, or by ammonium (NH_4^+). It is pH dependent, usually ascertained with cation replacement solutions buffered at pH 7 or 8.

Cellulose: A polysaccharide consisting of a long chain of glucose units.

Clod: A structureless mass of soil obtained by digging or plowing.

Colloid: Small particles in the nanometer to 1 µm range that can remain suspended in water and air.

Colluvium: Material moved down a slope mainly by gravity and deposited on the slope or at the foot of it.

Cryoturbation: Churning and mixing of soils by freezing and thawing of them.

Dalton: A unit of molecular mass that is equal to 1/12 of a carbon atom per mole.

Denitrification: The microbial transformation of nitrate or nitrite to NO_2 or N_2.

Density: A measure of the mass per volume, for example, Mg/m^3 or megagrams per cubic meter. *Particle density* and *bulk* (particles plus the air between particles) *density* are common soil terms.

Desertification: Soil degradation in dry areas that diminishes the capability of land to support viable ecosystems.

Detritus: The organic remains of dead organisms and organic material cast off from living organisms. Fresh and partially decayed leaves and other parts fallen from plants are commonly called *litter*, or they can be called (plant) *detritus*.

Dip slope: A slope that is along the surface of a tilted stratigraphic or bedding plane, such as the top of a sandstone or shale bed or layer, that was originally horizontal.

Ecosystem: A biological community and its environment. At the plant association scale it includes the soil and atmosphere around the plants.

Erodibility: Susceptibilities of soils, or other regolith materials, to be eroded by the actions of wind or water.

Erosion: Material loss from a landscape by gravitational mass displacement, transport by glacial ice, particulate transport by water or wind, or by weathering and solution loss.

Erosivity: The capabilities of wind, water, or ice to cause erosion.

Eukaryote: An organism with a cell nucleus of genetic material enclosed in a membrane.

Field capacity: The amount of water held in a freely drained soil about 2 days follow complete saturation.

Field water capacity: The amount of water held in a soil following drainage of excess water at atmospheric pressure (or -0.01 MPa).

Floodplain: The practically flat area adjacent to a stream that is occasionally, or frequently, flooded by water that flows over from the stream channel. It consists of alluvium deposited in a valley cut previously by the stream.

Grus: A structureless, unconsolidated mass of predominantly feldspar and quartz grains weathered from granitic rock.

Habitat: The environment occupied by living organisms.

Heat capacity: The amount of heat required to raise the temperature of a substance 1 K (or 1°C).

Heat, latent: Potential energy in a substance that can be released as heat at a phase transition, such as the freezing of water.

Heat, sensible: Heat that is related to the temperature of a substance; it can be felt by human senses.

Humus: Organic matter in soils that has been derived from the decay of plants and other organisms.

Hygroscopic water: Water adsorbed on surfaces and retained at very great (negative) potentials.

Landform: The form, or configuration, of land. Different kinds of landforms are distinguished by shapes and topographic positions. The naming of landforms commonly includes an interpretation of landform origin; for example, *floodplain* implies flooding of a flat, level area by water.

Landscape: A geographical unit consisting of one or more landforms. An *ecological landscape* includes the ecosystems in the landscape. An ecological landscape contains one or more contiguous ecological types.

Leaching: The transport of colloidal or dissolved substances through porous media. In soils, the direction is generally downward as water flows through them.

Lipids: Organic substances that are soluble in solvents such as ether and chloroform; they include fatty acids, waves, resins, and oils.

Loess: A deposit of aeolian dust that is predominantly silt and very fine sand.

Metabolism: The chemical processes in living organisms.

Meteoric water: Water from precipitation of moisture in the atmosphere, rather than water released from magma or rocks of Earth.

Mire: Wetland with soils that are perpetually wet.

Mitochondria: Membrane-bound intracellular bodies responsible for much oxidative respiration and generation of ATP.

Nitrification: The microbial oxidation of ammonium to nitrite (NO_2^-) and then to nitrate (NO_3^-).

Ped: A discrete aggregate of soil particles that is considered to be a unit of soil structure. Peds range in size from granules 1 to several mm in diameter to prisms and columns up to at least 100 mm in diameter.

Pedon: A soil unit with a horizontal area of about 1 to 10 m^2 that extends to a depth of about 2 m or to bedrock. It is the basic unit of soil description; smaller units do not encompass the range of properties that are used to differentiate different kinds of soils.

Peristaltic motion, or peristalsis: Longitudinal waves along the body of an organism caused by muscular contraction and relaxation.

Phyllosilicate: A mineral with planar sheets of silica tetrahedra. The sheets of silica tetrahedra are commonly combined with sheets of aluminum or magnesium octahedra to form layers. A combination of one each of tetrahedral and octahedral sheets is commonly designated a 1:1 layer silicate, and a combination of two tetrahedral and one octahedral sheets is designated a 2:1 layer silicate.

Polyphenols: Organic compounds composed of aromatic benzene rings with hydroxyl units attached to them.

Prokaryote: An organism lacking a cell nucleus of genetic material enclosed in a membrane.

Protista: A group of unicellular organisms consisting of protozoa.

Protoctista: A group of unicellular and simple multicellular eukaryotic organisms that are not considered to be plants, animals, or fungi.

Reclamation: Return of detrimentally altered land or soils to suitable land use.

Redox: An abbreviation of *reduction-oxidation* to indicate cycles of alternate oxidation and reduction.

Regolith: Unconsolidated and unlithified deposits or material at the surface of Earth.

Restoration of landscape features: The reclamation of viable ecological functions for degraded soils, biotic communities, and streams, and on a broader scale, watersheds.

Rheological properties: Properties of deformation and flow.

Runoff: Commonly means water that flows off over land rather than infiltrating into soils and flowing through the ground.

Salinization: Increase in surface soil salinity by natural processes or by inappropriate management. Commonly the salinity is caused by chloride and sulfate salts.

Saprolite: Soft, weathered bedrock that retains the fabric and structure of the parent rock.

Soil(s): Soil is defined differently for different disciplines:

Geology: Unconsolidated deposits of gravel, sand, silt, and clay just below the surface of Earth.

Pedology: Unconsolidated materials at the surface of Earth that are permanently or periodically occupied by meteoric water and gases and by living organisms.

Edaphalogy: Unconsolidated materials at the surface of Earth that support the growth of plants.

Engineering: All unconsolidated material above bedrock.

Cosmology: Unconsolidated material at the surface of Earth or an extraterrestrial body.

Soil degradation: Diminution of the capacity for a soil to grow plants, support natural biotic communities, and ameliorate hydrological processes such as infiltration, leaching of chemicals, and runoff of water.

Soil rehabilitation: The returning of a degraded soil to a desirable condition.

Soil restoration: The recovery of desired functions of a degraded soil.

Subsoil: An indefinite zone beneath the surface soil, or topsoil. It includes the B horizon if one is present.

Stratosphere: The next major layer of Earth's atmosphere above the troposphere.

Symbiosis: The cohabitation of dissimilar organisms in intimate association. It may or may not be beneficial to both of two cohabiting organisms.

Troposphere: The lowest major layer of Earth's atmosphere where the air is constantly in motion. It is about 11 km thick at middle latitudes and ranges from 7 km at the poles during summers to 20 km at the equator.

Watershed: The area above a stream location; the area from which water runs downslope to the stream location.

Water table: The upper surface of a groundwater body, where the water potential is equivalent to atmospheric pressure.

Weathering: Weathering is the physical and chemical alteration of rock or regolith in meteoric water, or at atmospheric pressure.

Wilting point: The soil water potential at which many plants wilt and die. It is assumed to be a potential of −1.5 MPa. Although many plants can survive in soils with much greater water stresses, most of them are merely surviving and not growing at the greater soil water stresses.

Appendix A: Extensive and Intensive Properties and Flow of Heat, Air, and Water in Soils

There are fundamental differences between heat and temperature, and between the amount of gas or water in a system and its pressure or potential, that are elucidated by acknowledging two classes of system properties—extensive and intensive properties. Heat and the mass of gas and water in a system are measures of amounts present in the system, whereas temperature, air pressure, and water potential can be measured independently from the amounts of heat, water, and air in a system. The former (heat and mass) are called *extensive*, and the latter are called *intensive* properties of a system (Table A.1). While addition of heat, gas, or water to a system will raise its temperature, pressure, or water potential, the values of the latter (temperature, etc.) can be measured without knowing the quantities of the former (heat, etc.).

The flow, or flux, of each substance depends on the relative magnitude of its intensive variable and the resistance to flow. Flux is the rate of flow per unit area; for example, a flow of 1 m^3/s through a column of soil 1 m_2 is a flux of 1 m/s. Laws of flow have the same form for each substance. Ohm's law is perhaps the most familiar. It may be stated that the flow of electricity, or electrical charge (I), is proportional to the potential gradient, or voltage differential along the path

Table A.1 Extensive and Intensive Properties of Common Substances

Substance	Extensive Properties		Intensive Properties		Law of Flow
	Name	SI Unit	Name	SI Unit	
Thermal energy	Heat	Joule (J)	Temperature	Kelvin (K) or Celsius (°C)	Fourier
Air	Molecules	Mole	Pressure	Pascal (Pa)	(Darcy, analogue)
Gas	Molecules	Mole	Partial pressure	Pascal (Pa)	Fick, diffusion
Water	Volume	Cubic meters	Pressure	Pascal (Pa)	Darcy
Electricity	Charge	Coulomb	Electrical potential	Volt (V)	Ohm

Note: Air is a mixture of gases. Each gas, or kind of gas, has only one kind of molecule, for example, N_2, O_2, or CO_2.

of flow (V), and inversely proportional to the resistance to flow (R); thus, I = V/R. In other laws, such as Darcy's, resistance is replaced by conductivity, which is the inverse of resistance. For example, in saturated soils, the flux of water is equal to the product of the conductivity (K_{sat}) and the pressure gradient, which is generally expressed as the difference in height or hydraulic head (δH) along the path of flow per distance or length along the path of flow (δL); thus, the flux, or rate of flow per unit area (q), is q = K_{sat} ($\delta H/\delta L$). The laws of Fick and Fourier (Table A.1) are analogous to those of Darcy in that the flux of a substance is proportional to the product of conductivity (inverse of resistance) and differences in concentration, pressure, or temperature gradients along the length of the path of flow.

Appendix B: Forms of Life

Half a century ago, there were only two kingdoms of life—plants and animals. Now any number up to eight or more kingdoms is recognized by different authorities. At the higher levels of taxonomy, classification has been in a state of flux for several decades as new molecular and chemical information used in classification continues to accumulate. Currently, there is no commonly agreed upon tree of life. Biologists agree, however, that life began with primitive microbes.

Biologists recognize eight levels of taxonomic classification (domain, kingdom, phylum, class, order, family, genus, and species). Seven kingdoms (first column) and a selection of phyla (third column) are presented in Table B.1. Many of the phyla are grouped or labeled by informal division, or group, names within each kingdom.

At the domain level, the primary division is between prokaryotes (Prokarya) with a chromosome (or nucleoid) free in cytoplasm and eukaryotes (Eukarya) with chromosomes (more than one per cell) enclosed in a nuclear membrane. The Prokarya are split between Archaebacteria (or Archaea) that lack peptidoglycan and Eubacteria, in which peptidoglycan is a major component of the cell wall. The cell differences allow Archaea to live and thrive in more extreme temperature, salinity, and alkalinity environments. Most Eubacteria thrive only at moderate temperatures (0–60°C) in nonsaline environments that are neither strongly alkaline nor strongly acidic.

Table B.1 Some Geoecologically Important Phyla in the Kingdoms of Life, Based on Genetic and Morphological Data

Kingdom	Division or Group	Phylum
Archaebacteria or Archaea		Euryarchaeota (methanogens and halophils)
		Crenarchaeota (thermoacidophils)
Eubacteria	Gracilicutes, gram (–)	Proteobacteria, purple bacteria (*Azotobacter, Desulfovirio, Nitrobacter, Nitrosomonas, Rhizobium, Thiobacillus*)
		Cyanobacteria (*Anabaena, Nostoc*, etc.)
	Tenericutes, wall-less	Aphragmabacteria (*Mycoplasma*, etc.)
	Firmicutes, gram (+)	Endospora (*Bacillus, Clostridium*, etc.)
		Actinobacteria (*Actinomyces, Frankia, Streptomyces*, etc.), including Coryneforms (*Arthrobacter, Corynebacteria*)
Protoctists	Amoebamorpha	Rhizopoda (amastigote amoeba, cellular slime molds)
		Foraminifera (protists including *Fusulina* and *Globigerina*)
	Alveolata	Ciliophora (ciliates, including *Paramecium* and *Vorticella*)
		Dinoflagellates, or dinomastigotes
		Oomycota (water molds, white rusts, downy mildews, including *Pythium, Phytophora*)
	Slime molds	Myxomycota (plasmodial slime molds)
Algae	Heterokonta	Euglenida (unicellular photoautotrophs, including *Euglena*)
		Bacillariophyta (protists with silica in their tests, diatoms)
		Chrysophyta (golden algae)
		Xanthophyta (yellow algae)
		Phaeophyta (brown algae)
	Isokonta	Cryptomonada (Cryptophyta, some photoautotrophs)
		Chlorophyta (green algae)
	Akonta	Rhodophyta (red algae)
	Slime nets	Labyrinthulata (*Labyrinthula*, only 8 species)
Fungi		Chytridiomycota (chitrids, protofungal microorganisms)
		Glomeromycota (arbuscular mycorrhiza, or endomycorrhiza)
		Zygomycota (many molds)
		Basidiomycota (arbutoid mycorrhiza and ectomycorrhiza)
		Ascomycota (ericoid mycorrhiza and most lichen fungi)

Table B.1 (*Continued*)   Some Geoecologically Important Phyla in the Kingdoms of Life, Based on Genetic and Morphological Data

Kingdom	Division or Group	Phylum
Animals	Parazoa	Porifera (sponges)
	Radiata	Coelenterata
		Centophora
	Acoelomates	Platyhelminthes (flatworms)
	Pseudocoelomates	Rotifera (rotifers, abundant in freshwater plankton)
		Nematoda (threadworms or roundworms)
	Coeleomates	
	Protostomes	Onychophora (velvet worms)
		Tardigrada (water bears)
		Annelida (earthworms, potworms)
		Mollusca
		Pelecypods (bivalve clams, mussels, and oysters)
		Gastropods (pulmonates such as snails and slugs)
		Cephalopods (squids, octopuses)
	Arthropoda	
		Chelicerata (spiders, mites)
		Myriapoda (centipedes, millipedes)
		Collembola (springtails)
		Hexapoda (insects)
		Crustacea (crayfish, woodlice)
	Deuterostomes	Echinodermata (sea stars, sea urchins)
	Chordata	Urochordata (ascidians, tunicates, or sea squirts)
		Craniata (fish, toads, salamanders, reptiles, birds, mammals)
Plants	Bryata (bryophytes)	Bryophyta (mosses)
		Hepatophyta (liverworts)
		Anthocerophyta (hornworts)
	Tracheata	Lycophyta (*Lycopodium, Selaginella*)
		Sphenophyta, horsetails (*Equisetum*)
		Pteridophyta, ferns (*Aspidotis, Pteridium,* etc.)
		Cycadophyta (*Zamia,* etc.)
		Gymnosperms (naked seeds)
		Coniferophyta, conifers (*Calocedrus, Pinus, Pseudotsuga*)
		Gnetophyta (*Ephedra,* etc.)
		Angiosperms (seeds enclosed in fruit), flowering plants
		Monocots (*Allium, Zigadenus,* etc.)
		Dicots (*Quercus, Ceanothus, Festuca,* etc.)

Note: Algae are in a group of photosynthetic organisms, without regard to taxonomic classification, other than they are not plants.

Microbiologists commonly call the prokaryote phyla of Margulis and Chapman (2009) kingdoms (Madigan and Martinko 2006) based on genetic differences, while Margulis and Chapman (2009) focus more on morphological differences. Madigan and Martinko (2006) recognize three kingdoms in the Archaebacteria, which they call Archaea, and 14 kingdoms in the Eubacteria, which they call Bacteria. Many biologists contend that the species concept is not applicable to prokaryotes.

Four kingdoms of commonly recognized groups can be split from the Eukarya, following Margulis and Chapman (2009): Protoctists, Plants, Fungi, and Animals. Plants and animals, which are familiar to everyone, develop from embryos. Fungi lack the flagella-like undulipodia that other eukaryotes have at some stage in their development (Margulis and Chapman 2009), and fungi reproduce by spores. Eukaryotes that are not plants, fungi, or animals are protoctists. Protoctists include both heterotrophs and photoautotrophs. Most of the photoautotrophic protoctists are called algae. Some protoctist phyla contain both photoautotrophic and heterotrophic species; for example, the dinoflagellates, or dinomastgotes, are about half heterotrophs and half photoautotrophs. Functionally, fungi are unicellular or multicellular absorptive heterotrophs; the animals are multicellular injestive heterotrophs; and the plants are multicellular photoautotrophs.

The second column in Table B.1 indicates informal groups of phyla. The third column is a list of many of the phyla described by Margulis and Chapman (2009) with some important examples of genera or other taxa within phyla in parentheses.

All animals in soils are bilaterally symmetrical. There are no Porifera, Coelenterata, Centophora, Echinodermata, or Urochordata in soils.

Within the plant kingdom, there is a major distinction between the kingdom and phyla levels that is very important in determining the functions of plants in geoecosystems. Vascular plants with lignified conducting systems and roots can have long stems, and their roots can extract water and nutrients from deep within soils. Nonvascular plants, which lack these features, are more dependent on dust and precipitation for sustenance. They are all very small, compared to trees and shrubs. They dry up when the atmosphere and ground surface dry out, while vascular plants remain active from stored water or from water that their roots extract from soils. Margulis and Chapman (2009) call the nonvascular plants Bryata and vascular plants Tracheata (Table B.1). The Bryata include mosses and liverworts, and the Tracheata include club mosses, horsetails, ferns, conifers, and flowering plants.

Appendix C: Notes on the Elemental Nutrition of Living Organisms

The elements required in relatively large amounts by living organisms are H and highly electronegative elements that form soluble oxyanions in water. These elements (H, C, N, O, P, and S) are shown in dark grey squares in Figure C.1. Those elements required in moderate amounts are alkali and alkaline earth cations (K, Ca, and Mg) in the two columns on the left in Figure C.1. Sodium (Na) is required by animals, but not by plants. Elements in the first row of transitional elements, from V to Zn, are required in small amounts, and Mo in the second row of transitional elements is required in minute amounts by Eubacteria, which fix N, which is then available to plants. Apparently, the crucial role of Mo in the nitrogenase of prokaryotes was once assumed by W, an element in the third row of transitional elements, below Mo. Vanadium replaces Mo in some N-fixing bacteria (for example, *Azotobacter*). The highly electronegative halogen elements (F, Cl, Br, and I) are either required by animals or highly beneficial to them. Boron is required in small amounts by plants, and Si is required by some plants and animals. The skeletal tests of diatoms, which are Protists (unicellular Protoctists), are composed of silica, and radiolarians have silica in their tests.

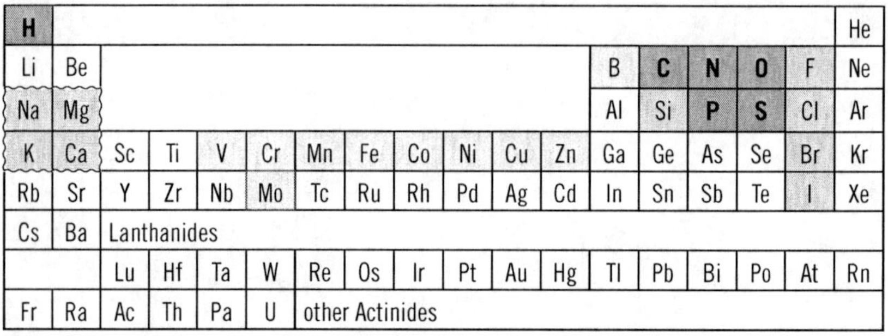

H																		He
Li	Be												B	**C**	**N**	**O**	F	Ne
Na	Mg												Al	Si	**P**	**S**	Cl	Ar
K	Ca	Sc	Ti	V	Cr	Mn	Fe	Co	Ni	Cu	Zn	Ga	Ge	As	Se	**Br**	Kr
Rb	Sr	Y	Zr	Nb	Mo	Tc	Ru	Rh	Pd	Ag	Cd	In	Sn	Sb	Te	I	Xe
Cs	Ba	Lanthanides															
		Lu	Hf	Ta	W	Re	Os	Ir	Pt	Au	Hg	Tl	Pb	Bi	Po	At	Rn
Fr	Ra	Ac	Th	Pa	U	other Actinides											

Figure C.1　Positions of essential chemical elements in a periodic table of the elements.

The toxic elements are more difficult to characterize than the essential ones. A reasonably successful approach is based on a combination of ionic potential and Lewis acid softness (Alexander et al. 2007, Appendix B). More simply, the most toxic elements are transition elements with filled d-orbitals (Cu, Zn, Ag, Cd, and Hg), and the less severely toxic are the oxyanions of Cr^{6+} and As^{5+}. Note that Cu (especially Cu^{2+}) and Zn, which are essential elements, are toxic in higher than essential concentrations. Antimony, thallium, and most of the elements with atomic numbers greater than that of Tl, from Pb on, are toxic, except Bi. The elements with even atomic numbers higher than 82 (Pb) are highly radioactive.

Appendix D: Rock Classification for Nongeologists

Many soils are derived from bedrock via parent material produced by the disintegration of *parent rock* (Soil Survey Staff 1993). The depth, structure, and chemistry of these soils are determined by the nature of the bedrock, among other things (Jenny 1980). These soil properties in turn affect the soil hydrology and plant survival and growth.

The bedrock properties that affect soil development are lithology and stratigraphy. Attitude (strike and dip of bedding, jointing, etc.) is important if the bedrock is stratified or has other parallel features. Bedrock attitude must be related to topography and slope gradient and aspect in order to interpret its affects on soils and ecosystems. Here the primary concern is the chemistry and mineral composition of the bedrock.

Geologists have identified an extensive array of rock types (Travis 1955). These rock types must be grouped into a manageable number of classes for soil and ecological landscape investigations. The scheme presented here is a guide, rather than a rigid system.

D.1 Major Grouping of Rock Types

Geologists have grouped rock types into three major divisions based on genesis: igneous, sedimentary, and metamorphic rocks (Bates

255

and Jackson 1980). Igneous rocks are those that have solidified from magma, or molten earth. Sedimentary rocks are formed by the consolidation of materials that have been deposited from suspension in air or water or precipitated from solution. Metamorphic rocks are formed by the alteration of preexisting rocks, due to substantial changes in the temperature, pressure or stress, or chemical environments. As with any genetic classification, there are some ambiguous cases. For example, metasomites and migmatites (Travis 1955) are rocks that are only partially altered by chemical processes or remelting. Two of the many good general books on petrography and petrology are those of Williams et al. (1982) and Jackson (1970). There are many more specific books on only one or two of the major rock types, for example, Middlemost (1985), Tucker (1991), and Hyndman (1985).

A relatively definitive grouping of rock types has been made for igneous rocks. The sedimentary and metamorphic counterparts of some chemical or biological precipitates, such as $CaCO_3$ (limestone and marble), are more practically grouped across major divisions than with other rock types within the same major divisions; for example, soils derived from limestone, a sedimentary rock, are more similar to marble, a metamorphic rock, than to chert, another sedimentary rock that is chemically or biologically precipitated silica (SiO_2).

D.2 Igneous Rocks

The primary division of igneous rocks is based on crystal grain size distributions that depend largely on depth of emplacement and rate of cooling and solidification. Igneous rocks formed by slow cooling of magma deep within Earth are called plutonic rocks. And those formed by more rapid cooling at the surface of Earth are called volcanic rocks. There are grey zones, however, between where rocks have formed by intermediate rates of cooling and solidification at relatively shallow depths. Petrologists have classified lamprophyres, which are groups of dike rocks, separately from plutonic and volcanic rocks (Streckeisen 1979), but those rocks are of no concern to the vast majority of ecological land managers.

The primary subdivision of igneous rocks is made more operational (based on properties rather than on genesis) by a differentiation between phaneritic (coarse-grained) and aphanitic (fine-grained) rocks (Travis 1955). Plutonic rocks consistently have larger (phaneritic) crystals, due to slower cooling and solidification, and volcanic rocks generally have smaller crystals in a microcrystalline or glassy ground-mass (aphanitic texture).

D.2.1 Plutonic Rocks

The subdivision of plutonic (phaneritic) rocks is based primarily on mineralogy. The key mineral classes are quartz, alkali (K and Na) feldspars, nonalkali (Ca $\geq$ Na) plagioclase feldspars, feldspathoids, and mafic minerals (especially hornblende, pyroxene, and olivine). An internationally accepted classification of plutonic rocks is based on two complementary triangles (LeMaitre 1989): one for an alkalic series of rocks and another for a calc-alkalic series. Because the great majority of plutonic rocks are in the calc-alkalic series (Jackson 1970), let us ignore the alkalic series. The ultramafic igneous rocks, those with low silica (SiO_2 < about 45 or 48%) concentrations are classified on a separate triangle, with olivine, orthopyroxene, and clinopyroxene apices. A simple table will suffice for most soil and ecological land-scape investigations involving igneous rocks (Table D.1).

Table D.1 Common Kinds of Igneous Rocks in the Calc-Alkaline Series, and a Few Less Common Rocks, such as Syenite and Ultramafite

Crystal Grain Size	Quartz[a] (%)	Silica Content (%)				
		70	61	52	45	
Fine (aphanitic)	20–60	Rhyolite	Dacite	—	—	—
	5–20	—	Quartz latite	Andesite	—	—
	0–5	—	Trachyte	Latite/Andesite	Basalt	Ultramafitite[b]
Fine, phaneritic					Diabase[c]	
Coarse (phaneritic)	20–60	Granite	Granodiorite/tonalite	—	—	
	5–20	—	Quartz Monzonite	Quartz diorite	—	—
	0–5	—	Syenite	Monzonite	Diorite/Gabbro	Peridotite

[a] Including chemically identical cristobalite and tridymite in volcanic rocks, particularly in andesite and basalt.

[b] Presumably komatiite; picrite has silica < 45% but lacks the very dark ultramafic color.

[c] The preferred name is microgabbro (Streckeisen 1979), but diabase and dolerite are more widely utilized names.

D.2.2 *Volcanic Rocks*

The subdivision of volcanic (aphanitic) rocks is also based on mineralogy (LeMaitre 1989), with the same key mineral classes as in the triangles for the classification of plutonic rocks, although cristobalite and tridymite are included with quartz in the classification of volcanic rocks. Again, we can ignore the alkalic series without missing many of the volcanic rocks that are exposed at the surface of Earth (Jackson 1970). Ultramafic volcanic rocks are uncommon.

Some volcanic rocks with similar chemical composition may have textural differences that are significant for soil and ecological investigations. The glassy rocks, which form by very rapid cooling of magma, are an example. Obsidian, which is the most common glassy rock, is generally the chemical equivalent of rhyolite.

D.3 Sedimentary Rocks

Sedimentary rocks are conveniently divided into clastic sediments produced by erosion, transport, and settling of discrete particles and nonclastic sediments produced by chemical and biological deposition. These two classes, clastic sediments and chemical-biological deposits, are not entirely unequivocal. Many sediments are mixtures and many rocks contain both kinds of sediments.

D.3.1 *Clastic Sedimentary Rocks*

Events leading to the formation of clastic sediments are erosion of particles, or regolith; particle transport by gravity, wind, water, or ice; and settling of particles from water or air. After deposition, sediments transformed by compaction, chemical alteration, and cementation are lithified to produce sedimentary rocks.

The clastic sedimentary rock properties that are most important in a classification for soil and ecological investigations are texture, particularly grain size distribution, particle composition, matrix (finer material < 0.062 mm surrounding coarser particles), and cement. Three main clastic sedimentary rock classes are distinguished, based on predominant grain sizes: (1) *conglomerate* with pebbles, cobbles, and boulders; (2) *sandstones* with sand; and (3) *mudstones* and *shales* with silt and clay.

Coarse-grained rocks are sometimes called *rudites* (a seldom used term) to include both conglomerates with rounded particles and breccias with angular particles. The particles are generally rock fragments. Landslide deposits and glacial till are commonly unsorted sediments that are called *diamicton*. Alluvial deposits are generally sorted, separating the finer from the coarser particles to yield conglomerates with negligible silt and clay. Lithified diamicton is called diamictite, or tillite if the origin is known to be till.

Fine-grained (sand and smaller particles) sedimentary rocks are commonly called *arenite* if the nonsandy matrix (particles < 0.062 mm) is < 15%, *wacke* if the matrix is 15 to 75%, and *lutite* (mudstone, siltstone, or claystone) if the nonsandy matrix is > 75%. Mudstone has a mixture of silt and clay size particles. Lutite with fissile (splitting) properties is called shale.

Sandstones are named according to the amount of nonsand matrix and the mineralogical composition of the sand grains (Table D.2). The common sandstone cements are silica (or microcrystalline quartz), calcite, iron oxides (especially goethite and hematite), and clay. Gypsum and siderite are less common cements.

Mudstones and shales have predominantly quartz and feldspars in the silt and clay minerals of the finer fractions. Whereas kaolinite and smectite are the common end products of weathering in acidic and alkaline environments, kaolinite and illite are the common products of the diagenetic alteration to yield shale. Metamorphism produces sericite and chlorite.

Table D.2 Sandstone (Arenite)[a] Classification

Dominant Sand Grain Composition	Matrix Content (particles < 0.062 mm)	
	0–15%	15–75%
Quartz	Quartzite[b] or quartzarenite	Quartz wacke
Feldspar	Arkose or arkosic arenite	Arkosic wacke or feldspathic greywacke
Rock fragments	Litharenite or lithic arenite	Lithic greywacke

[a] Arenite is a petrographic name for sandstone.
[b] Quartzite is a confusing name that does not distinguish between sedimentary and metamorphic quartzites. Sedimentary quartzite is sometimes called orthoquartzite, but quartzarenite is more consistent with the other terms in this table.

D.3.2 Nonclastic Sedimentary Rocks

The common ones are limestone and chert. Coal, Precambrian iron-stones, phosphorites, and evaporites are of greater economic importance but much less extensive.

Limestone is calcium carbonate ($CaCO_3$), predominantly calcite and less aragonite. Many old limestones have been partially or completely dolomitized by the addition of Mg. Pure dolomite has molar equivalent amounts of Ca and Mg; it is $CaMg(CO_3)_2$.

Chert results from the biochemical precipitation of silica (SiO_2) as opal and its diagenetic transformation via cristobalite and tridymite to quartz. Varieties are named for their colors; for example, grey chert is called flint and red chert is called jasper.

Evaporites contain many minerals, but halite ($NaCl$), gypsum ($CaSO_4 \cdot 2H_2O$), and anhydrite ($CaSO_4$) are the common ones. All are soluble in water.

D.3.3 Pyroclastic Rocks

These may be treated as either igneous or sedimentary rocks. They are those volcaniclastic rocks that formed from magma ejected from volcanic vents (igneous process) and were then deposited by falling or settling from the atmosphere on land or in water (sedimentary process). The main differentiating characteristics of pyroclastic rocks are particle size and lithification.

Fine-grained (particles < 2 mm) deposits are *ash* and *tuff*, which is lithified ash. They are generally siliceous, with mafic ash being much less common.

Cinders are pebble size pyroclasts. Siliceous pyroclasts of comparable size and form are called *pumice*.

Large pyroclasts (particles > 64 mm) are called *blocks* if solidified before ejection and *bombs* if the magma solidifies after being ejected from a vent. Pyroclastic deposits with predominantly blocks is called volcanic *breccia*. With sufficient lithified matrix, the rocks are called *tuff-breccia*. Volcanic breccias may form by the direct fall of blocks from the atmosphere or indirectly by falling on water or snow and settling from water. Volcanic mudflow deposits are generally tuff-breccia, commonly called *lahar*.

D.4 Metamorphic Rocks

Metamorphic rocks are a very complex group. Their classification is less systematic than that of igneous and sedimentary rocks. It is convenient and appropriate for soil and ecological investigations to separate metamorphic rocks with specific mineral composition (Table D.3) from the others (Table D.4).

Only four kinds of metamorphic rock with specific mineral composition are sufficiently extensive to warrant listing in Table D.3. *Quartzite* and *marble* are generally massive, without appreciable linear orientation of grains. *Serpentinite* commonly appears massive, although crystals are oriented in veins and locally on a microscopic scale, but the sheared version with polished surfaces is the most familiar to people. *Amphibolite* generally has some schistosity, but more like a gneiss than a schist.

Table D.3 Metamorphic Rocks with Specific Mineral Composition, Generally Only One or Two Minerals

Metamorphic Rock Type	Igneous or Sedimentary Precursor	Mineral Composition
Metaquartzite	Quartzarenite	Quartz
Marble	Limestone	Calcite or dolomite
Amphibolite		Mostly amphibole and plagioclase
Ortho-amphibolite	Igneous rock	
Para-amphibolite	Metamorphic rock	
Serpentinite	Peridotite	Serpentine
Coal[a]	Lignite	Peat (organic matter)

Note: Skarn (or tactite) and eclogite might be added, but they are much less extensive than the other rock types.

[a] Bituminous coal is a lower-grade and anthracite is a higher-grade metamorphic product.

Table D.4 Metamorphic Rocks without Specific Mineral Composition

Relative Grain Size	Regional Metamorphism			Contact Metamorphism
	Foliation or Linear Alignment of Grains			
	Minimal	Moderate	Maximal	Homogranular[a]
Fine	Slate	Phyllite	Schist	Hornfels
Coarse	Granofels	Gneiss	Schist	(Granulite)[b]

Note: Skarn (or tactite) might be added in the table, but it is much less extensive than the other rock types.

[a] Crystal grains of more or less equal size and any elongated grains are randomly oriented.

[b] *Granulite* has different meanings in different contexts and may not be a very useful term.

Metamorphic rocks lacking specific mineral composition are placed in one of three categories, based on the degree of schistosity (Fettes and Desmon 2007, International Union of Geological Sciences (IUGS) recommendations): (1) *schist* with fine foliation, (2) *gneiss* with coarse foliation, and (3) *granofels* lacking foliation. These names based on specific structure may be given modifiers, for example, mica-quartz schist.

Rocks may be separated into two or more classes based on the mode of metamorphism. The most common mode classes are regional and contact metamorphism. Regional metamorphism is due to both pressure and temperature, resulting in rocks that are generally foliated or have linearly oriented grains. Contact metamorphism is due to heat, commonly heat from an igneous intrusion, resulting in rocks that are generally homogranular (all crystals nearly the same size) without significant lineation. Rocks altered by regional metamorphism can be subdivided by degree of foliation (Table D.4).

Intensive shearing produces rocks of great petrographic interest but insufficiently extensive to be significant in soil and ecological investigations. Most notable is mylonite, which is a foliated, medium- or fine-grained to flinty or glassy, streaky-looking rock. Fault breccia is called cataclasite.

Metamorphic petrologists commonly use the kinds of minerals in rocks to infer the pressure and temperature of alteration, or metamorphism. The main low-pressure *metamorphic facies* from low to high temperature are zeolite, greenschist, amphibolite, and hornfels or granulite. The blueschist facies is produced at relatively low temperature and high pressure. Blueschist, which contains an amphibole mineral called glaucophane, has been cited as an indicator of subduction of the oceanic Franciscan complex beneath continental crust, and prompt exhumation.

D.5 Perspective for Soil and Ecological Landscape Investigations

A rock classification for soil and ecological landscape investigations and management should be based primarily on parent rock properties that affect soil texture, structure, depth, hydrology, fertility, and slope stability.

D.5.1 *Igneous Rocks*

The important differentiating features of the igneous rocks are (1) grain size, which has pronounced effects on soil texture and hydrology, and (2) chemistry and mineralogy, which has pronounced effects on fertility, soil texture, and clay mineralogy.

Two classes are proposed for rock grain size (phaneritic and aphanitic) and three for chemistry and mineralogy, which combined yield 6 classes of igneous rocks (Table D.5), compared to 10 combined classes in Table D.1. The three chemical classes, based on silica content, are siliceous (SiO_2 > about 56 to 61%), mafic (intermediate SiO_2 content), and ultramafic (SiO_2 < 45%) rocks. Igneous rock classification based on mineralogy yields similar groups, but with mineral rather than chemical criteria.

The siliceous phaneritic, or plutonic, rocks generally weather to a coarse sandy to very fine gravelly grus in tropical and temperate climates. This kind of weathering has been attributed to the hydration and expansion of mica, predominantly biotite (Wahrhaftig 1965). The grus-forming rocks (granite, quartz monzonite, granodiorite, etc.) are commonly called granitic rocks. One of the distinctive features of soils in grus is that they are highly susceptible to gully erosion. Soils developed from peridotite and other ultramafic plutonic rocks are infertile, due mainly to very high Mg and low Ca contents (Alexander et al. 2007). Intermediate between the siliceous and ultramafic rocks are the mafic rocks—hornblende diorite and gabbro. Diorite is a grus-forming rock if it has enough mica or, otherwise, a mafic rock. Some gabbro yields infertile soils (Hunter and Horenstein 1992). Thus, although Table D.1 is a guide to grouping igneous parent rock classes of Table D.5, the final criteria for ecological landscape investigations

Table D.5 Rock Guide to Soil and Substratum Features Very Important for Soil and Ecological Landscape Investigations of Terrain with Igneous Rocks

Rock Grain Size Class	Rock Chemistry Class		
	Silicic	Mafic	Ultramafic
Aphanitic	Soil fertility less than maximum	Fertile soils	—
	No grus	Vertically jointed bedrock	
Phaneritic	Grus-forming rock	Fertile soils	Infertile soils
	Unstable or erodible soil	No grus	

should be the soils, their fertility, and their natural plant communities or responses to management.

Most of the fine-grained, or aphanitic, igneous rocks are volcanic rocks. The siliceous rocks (rhyolite, dacite, etc.) are generally less fertile than the mafic rocks (andesite, basalt, etc.). Most ultramafic volcanic rocks (komatiites, etc.) are Precambrian, and very little of those are currently exposed at the surface of Earth. Basalt generally has a higher Ca content than other calc-alkalic volcanic rocks, but some rocks near ultramafic composition classified as basalt, based on color or mineralogy, have high Mg and relatively low Ca contents and yield infertile soils.

The siliceous volcanic rocks are not grus formers, a key criterion in landscape investigations. A major difference between mafic aphanitic and plutonic rocks is that the former, particularly basalt, have vertical joints that are conduits for water; thus, the hydrology is much different. The most important thing to remember is that the petrological class limits do not correspond exactly to the most practical group limits for ecological landscape investigations. Any delineation for ecological landscape investigations based on geologic maps must be checked in the field.

D.5.2 Sedimentary Rocks

The major classes of clastic sedimentary rock—conglomerate (or rudite), sandstone, and lutite (mudstone and shale)—are helpful, but not necessarily the most useful classes for ecological landscape investigations. Soil types from soft or slightly hard mudstone or shale are dependent on the clay minerals in the lutite, whereas soils developed from hard shale may be similar to those of some pelitic metamorphic rocks (phyllite and slate). Soils from mudstone with smectitic clays are generally expansive and, where they are on slopes, generally have slope instability or erosion problems under management. Sandstone classification is based on the composition of the sand grains and the silt or clay matrix content (Table D.2), but the cement may have as much, or more, influence on soil development. Sandstones with carbonate cement commonly form sandy soils, and greywackes generally do not. Soils most likely to be sandy are those with quartz sand grains, which are resistant to weathering, and $CaCO_3$ or a more soluble cement.

The major nonclastic sedimentary rock classes, limestone (and dolomite) and chert, are convenient for ecological landscape investigations too. Marl and chalk should be differentiated from hard limestone. Evaporites are a less extensive and very diverse class.

D.5.3 Metamorphic Rocks

Metaquartzite, marble, and serpentinite each yield distinctive soils, commonly with vegetation that is distinctly different from that on other soils. Soils on metaquartzite may, or may not, be similar to those of quartz arenite and chert. Hard limestone and marble generally have similar soils that need not be differentiated in ecological landscape investigations, even though one is a sedimentary and the other a metamorphic rock. Soils on serpentinite are dissimilar from those on any other rock type, but their vegetation is practically indistinguishable from that of soils on peridotite. Soils on amphibolite may be similar to those on mafic igneous rocks or to soils of some other unfoliated, or slightly foliated, metamorphic rocks; thus, it is not necessary in all areas to have a separate map unit for soils or ecological landscapes on amphibolite only.

Generally, five or less groups of soil or ecological landscape units are necessary for the slate-phyllite-schist-gneiss-granulite sequence of metamorphic rocks. Soils on granulite with little foliation may be sufficiently similar to those on granofels or hornfels to be in the same soil or ecological landscape units. Some gneiss may weather to grus and have soils similar to those of granitic rocks. The admonition that the final criterion for separating map or landscape units is the soil and natural vegetation and its productivity is particularly important in investigations areas with the metamorphic rocks of Table D.4; the petrological classes or class limits are not necessarily the ones that are important in soil and ecological landscape investigations.

Appendix E: Naming the Suborders, Great Groups, and Subgroups in *Soil Taxonomy*

The U.S. *Soil Taxonomy* is a formal system. It was initiated in the United States during the middle of the twentieth century and has become an international endeavor that is supervised by the Natural Resources Conservation Service (NRCS) of the U.S. Department of Agriculture (Soil Survey Staff 1975, 1999). Only names of taxa defined in publications of the NRCS are acceptable. Revisions of *Soil Taxonomy* are published in *Keys to Soil Taxonomy* every few years. The 11th edition of *Keys to Soil Taxonomy* was published in 2010 (Soil Survey Staff 2010).

The orders of *Soil Taxonomy* are described in Chapter 6. Naming suborders, great groups, and subgroups is a formal procedure. Formative elements for suborders and for great groups are added as prefixes to formative endings of a soil order; then the great group name is preceded by an adjective designating the subgroup. Formative endings for the 12 orders are listed in Table 6.4, formative elements for suborders and great groups are listed in Table E.1, and adjectives to designate a subgroup are listed in Table E.2.

Table E.1   Formative Elements Utilized in Naming Suborders and Great Groups

Element	Origin	Brief Explanation and Meaning in Soil Taxonomy
Acr	Gr. *akros*, at the end	Weathered completely
Agr	L. *ager*, field	Having an agric horizon
Al	*Aluminium* or *aluminum*	Aluminum present, without iron
Alb	L. *albus*, white	Bleached by leaching of iron
Anhy	Gr. *anydros*, waterless	Anhyrous conditions
Anthr	Gr. *anthropos*, man	Having an anthropic epipedon
Aqu	L. *aqua*, water	Wet long enough to form redoximorphic features
Ar	L. *arare*, to plow	Mixed artificially, not by natural processes
Arg	L. *argilla*, clay	Having an argillic horizon
Bor	Gr. *boreas*, northern	A cold soil
Calc	L. *calcis*, lime	Having a calcic horizon
Camb	L. *cambiare*, to exchange	Having a cambic horizon
Cry	Gr. *kryos*, icy cold	Cryic soil temperature regime
Dur	L. *durus*, hard	A duripan present
Dys, dystr	Gr. *dys*, ill or bad	Low basic cation content
Endo	Gr. *endon*, within	Saturation from below
Epi	Gr. *epi*, at, on, over	Saturation from above, as in a perched water table
Eu	Gr. *eu*, good or well	High basic cation status
Ferr	L. *ferrum*, iron	Iron-cemented nodules present
Fibr	L. *fibra*, fiber	Organic material with a high fiber content
Fluv	L. *fluvus*, river	Fluvial stratification that affects vertical sequence of soil properties
Fol	L. *folia*, leaf	Well-drained organic soil formed by accumulation of leaves
Frag	L. *fragilis*, brittle	A fragipan present
Fragloss	Fra(g) + gloss	
Fulv	L. *fulvus*, tawny	Dark-colored epipedon, not melanic, high organic matter content
Glac	L. *glacies*, ice	With a glacic layer
Gloss	Gr. *glossa*, tongue	A glossic horizon present
Gyps	L. *gypsum*, gypsum	A gypsic horizon present
Hal	Gr. *hals*, salt	High exchangeable sodium concentration
Hapl	Gr. *haplous*, simple	No special features
Hem	Gr. *hemi*, half	Organic material with moderate fiber content
Hum	L. *humus*, earth	High humus, or organic matter, concentration
Hydr	Gr. *hydor*, water	Continuously under water
Kandi	*kandite*, kaolinite group	A kandic horizon present
Kanhapl	Kan(di) + hapl	Lacking depth requirements of a kandic horizon
Luv	Gr. *louo*, to wash	Contains humilluvic material, illuviated organic matter

Table E.1 (*Continued*)   Formative Elements Utilized in Naming Suborders and Great Groups

Element	Origin	Brief Explanation and Meaning in *Soil Taxonomy*
Med	L. *media*, middle	Moderate temperature, neither hot tropical nor very cold
Melan	Gr. *melas*, black	A melanic epipedon present
Natr	Gr. *nitron*, niter	A natric horizon present
Ochr	Gr. *ochros*, pale	An ochric epipedon present
Orth	Gr. *orthos*, true	No special features
Pale	Gr. *paleos*, old	Advanced soil development
Per	L. *per*, thorougly	A perudic soil moisture regime
Pergel	Per + gel(i)	
Petr	Gr. *petra*, rock	A hardpan present
Plac	Gr. *plax*, flat stone	A placic horizon present
Plagg	Ger. *plaggen*, sod	A plaggen epipedon present
Plinth	Gr. *plinthos*, brick	Plinthite present
Psamm	Gr. *psammos*, sand	Sandy
Quartz	Ger. *quarz*, quartz	Very high quartz content in sand fractions
Rend	Rus. *rendzina*, lime-rich soil	Calcium carbonate in a mollic epipedon or just below it
Rhod	Gr. *rhodon*, rose	Reddish colors of high chroma and a dry value nearly equal to the moist value
Sal	L. *sal*, salt	A salic horizon present
Sapr	Gr. *sapros*, rotten	Decomposed organic material with a low fiber content
Sombr	Fr. *sombre*, dark	A sombric horizon present
Sphag	Gr. *sphagnos*, bog	Fibric organic material containing mostly *Sphagnum* spp.
Sulfi	*Sulfide*	Sulfidic materials present
Sulfo	L. *sulfur*, sulfur	A sulfuric or a sufidic horizon present
Torr	L. *torridus*, hot and dry	Very dry soil
Trop	Gr. *tropikos*, of the solstice	Equable soil temperatures, lacking annual extremes
Turb	L. *turbo*, whirlwind	Evidence of cryoturbation
Ud	L. *udus*, humid	Soil moist through most of growing season, udic soil moisture regime
Umbr	L. *umbra*, shade	Presence of umbric epipedon
Ust	L. *ustus*, burnt	Soil periodically dry over > 50% of growing season, ustic SMR
Verm	L. *vermes*, worm	Abundant worm holes or worm casts or filled animal burrows
Vitr	L. *vitrum*, glass	Containing volcanic glass, at least 5% in sand and coarse silt fractions
Xer	Gr. *xeros*, dry	Soil dry during summer, xeric soil moisture regime

Table E.2 Subgroup Adjectives Preceding Great Group Names

Subgroup Name	Origin	Meaning
Abruptic	L. *abruptum*, turn off	Abrupt texture change, vertically
Aeric	Gr. *aer*, air	Aerated ephemerally saturated soil
Albic	L. *albus*, white	With an albic horizon
Alic	*Aluminium* or *aluminum*	
Andic		High amorphous silicate content
Anionic	L. *anion*, (thing) going up	High anion adsorption capacity
Anthraquic	Anthr(a) + aquic	Artificially flooded, as in rice paddies
Anthropic	Gr. *anthropos*, man	With an anthropic epopedon
Aquic	L. *aqua*, water	Ephemerally saturated soil
Arenic	L. *arena*, sand	Sandy epipedon
Argic	L. *argilla*, clay	An argillic horizon
Aridic	L. *aridus*, dry	An aridic soil moisture regime
Calci	L. *calcis*, lime	A calcic horizon
Chromic	Gr. *chroma*, color	High color value
Cumulic	L. *cumulus*, heap	Thickened epipedon
Duric	L. *durus*, hard	Silica cementation
Durinodic	Duri + nodic	Nodules of silica-cemented soil
Dystric	Gr. *dys*, bad; *trophia*, nutrition	
Entic	*Recent*	Minimal development
Eutric	Gr. *eu*, good; *trophia*, nutrition	
Fibric	L. *fibra*, fiber	
Fragic	L. *fragilis*, brittle	Fragic (brittle when moist) soil properties
Glacic	L. *glacies*, ice	
Glossic	Gr. *glossa*, tongue	With a glossic horizon
Grossarenic	L. *grossus*, thick; L. *arena*, sand	Thick sandy epipedon
Gypsic	L. *gypsum*, gypsum	
Halic	Gr. *hals*, salt	
Haplic	Gr. *haplous*, simple	
Hemic	Gr. *hemi*, half	
Histic	Gr. *histos*, tissue	With a histic epipedon
Humic	L. *humus*, earth	
Hydric	Gr. *hydor*, water	Continually saturated with water
Lamellic	L. *lamella*, a thin layer	Thin, horizontal layers with illuvial clay
Leptic	Gr. *leptos*, thin	Thin soil, moderately deep
Limnic	Gr. *limn*, lake	
Lithic	Gr. *lithos*, stone	A shallow lithic contact
Mollic	L. *mollis*, soft	Some characteristics of a mollic epipedon
Natric	Sp. *natron*, sodium carbonate	With a natric horizon
Nitric	Gr. *nitron*, niter	High nitrate concentration
Ombraquic	Gr. *ombros*, shower of rain	Reduced surface soil over oxidized subsoil

Table E.2 (*Continued*)   Subgroup Adjectives Preceding Great Group Names

Subgroup Name	Origin	Meaning
Oxic	*Oxygen*	
Oxyaquic	Oxy + aquic	Saturation without appreciable reduction
Pachic	Gr. *pachys*, thick	Thickened epipedon on slopes
Pergelic	Per + gelic	Subfreezing soil
Petrocalcic	Gr. *petra*, rock; L. *calcis*, lime	A petrocalcic horizon
Petroferric	Gr. *petra*, rock; L. *ferrum*, iron	Indurated by cementation with iron compounds
Petronodic	Gr. *petra*, rock; L. *nodus*, a knot	Indurated nodules (L. *nodulus*, a little knot)
Plinthic	Gr. *plinthos*, brick	Plinthite, iron concentrations that harden on drying
Psammentic	Gr. *psammos*, sand	Sandy soil
Rhodic	Gr. *rhodon*, rose	
Ruptic-	L. *ruptum*, broken	Irregular thickness or discontinuous
Salidic	L. *sal*, salt	Salty soil, with a salic horizon
Sapric	Gr. *sapros*, rotten	Well-decomposed organic material
Sodic	*Sodium*	High sodium concentration
Sombric	Fr. *sombre*, somber	With a sombric horizon
Spodic	Gr. *spodos*, wood ash	
Sphagnic	Gr. *sphagnos*, bog	
Sulfic	*Sulfide*	
Sulfuric	*Sulfuric* (acid)	
Thaptic-	Gr. *thapto*, buried	A buried horizon or soil
Terric	L. *terra*, earth	A mineral substratum in an organic soil
Typic	L. *typicus*, Gr. *typikos*, typical	
Udic	L. *udus*, humid	
Umbric	L. *umbra*, shade	
Vermic	L. *vermes*, worm	
Vertic	L. *verto*, turn	Cracking clay, homogenized vertically
Xanthic	Gr. *xanthos*, yellow	Yellowish color
Xeric	Gr. *xeros*, dry	Nearly dry enough for a xeric soil moisture regime

Example for Naming a Suborder, Great Group, and Subgroup of an Alfisol

> Formative ending for the order, *alf*
> Formative prefix for a suborder, *aqu*
> Suborder name, *Aqualf*
> Formative prefix for a great group, *alb*
> Great group name, *Albaqualf*
> Adjective for a subgroup, *aeric*
> Subgroup name, *Aeric Albaqualf*

Only names that have been approved and appear in NRCS publications for *Soil Taxonomy* are acceptable; for example, there are Aqualfs, Cryalfs, Ustalfs, Xeralfs, and Udalfs, but no Albalfs—the *alb* formative prefix is utilized at the great group level, rather than at the suborder level in the Alfisol order, as in the Albaqualfs.

Appendix F: Taxonomic Names of the Plants

Plants are identified by common names in the text. Because different common names are used in different areas, taxonomic names corresponding to the common names used in the text are listed in Table F.1. Botanical authorities are important for plant taxonomists, but for others, adding the authorities to taxonomic names is more confusing than helpful. Botanical authorities for the taxonomic names can be found on the plants.usda.gov website.

Table F.1 Taxonomic Names of Plants Listed by Common Name

Common Name	Life-Form	Taxonomic Name
Bitterbrush	Shrub	*Purshia tridentata*
Bluegrass	Grass	*Poa secunda*
Brome	Grass	*Bromus* spp.
Soft		*B. hordeaceus*
Smooth		*B. inermis*
Ceanothus, prostrate	Shrub	*Ceanothus prostratus*
Cheatgrass	Grass	*Bromus tectorum*
Chinquapin, bush	Shrub	*Chysolepis sempervirens*
Downingia, false	Forb	*Porterella carnosa*
Draba, cushion	Forb	*Draba* sp.
Fescue, Idaho	Grass	*Festuca idahoensis*
Fir	Tree	*Abies* spp.
Red		*A. magnifica*
White		*A. concolor*
Greasewood, black	Shrub	*Sarcobatus vermiculatus*
Hairgrass, tufted	Grass	*Deschampsia cespitosa*
Heather, white	Subshrub	*Casiope mertensiana*
Incense—cedar	Tree	*Calocedrus decurrens*
Joint-fir, Nevada	Shrub	*Ephedra nevadensis*
Juniper	Tree	*Juniperus* spp.
Western		*J. occidentalus*
Lupine	Forb	*Lupinus* spp.
Dwarf		*L. lepidus*
Velvet or wooly leaf		*L. leucophyllus*
Mahogany, leatherleaf	Shrub	*Cercocarpus ledifolius*
Mountain pride	Subshrub	*Penstemon newberryi*
Manzanita	Shrub	*Arctostaphylos* spp.
Common		*A. manzanita*
Greenleaf		*A. patula*
Mat (pinemat)		*A. nevadensis*
Mendora, spiny	Shrub	*Mendora spinescens*
Muhly, mat	Grass	*Muhlenbergia richardsonis*
Needlegrass	Grass	*Achnatherum, Nassella, Stipa* spp.
Oak	Tree	*Quercus* spp.
Black		*Q. kelloggii*
Blue		*Q. douglasii*
Interior live		*Q. wizlizenii*
Oat, wild	Grass	*Avena barbata, A. fatua*
Phlox, cushion	Forb	*Phlox condensata*

Table F.1 (Continued)   Taxonomic Names of Plants Listed by Common Name

Common Name	Life-Form	Taxonomic Name
Pine	Tree	*Pinus* spp.
Gray		*P. sabiniana*
Jeffrey		*P. jeffreyi*
Lodgepole		*P. contorta*
Ponderosa		*P. ponderosa*
Sugar		*P. lambertiana*
Yellow		*P. jeffreyi, P. ponderosa*
Whitebark		*P. albicaulis*
Piñon	Tree	*Pinus monophylla*
Pride, mountain	Subshrub	*Penstemon newberryi*
Ricegrass, Indian	Grass	*Achnatherum hymenoides*
Rush	Graminoid	*Juncus* spp.
Sagebrush, big	Shrub	*Artemisia* spp.
Big		*A. tridentata*
Bud		*A. spinescens*
Low		*A. arbuscula*
Silver		*A. cana*
Saltbush, big	Shrub	*Atriplex lentiformis*
Shadscale	Shrub	*Atriplex confertifolia*
Snowberry,	Shrub	*Symphoricarpos* spp.
Creeping		*S. mollis*
Roundleaf		*S. rotundifolius*
Spikerush	Graminoid	*Eleocharis* spp.
Wheatgrass	Grass	*Elymus spicatus*
Whitethorn, mountain	Shrub	*Ceanothus cordulatus*

References

Aerts, R. 1997. Climate, leaf litter chemistry, and leaf litter decomposition in terrestrial ecosystems a triangular relationship. *Oilos* 79: 439–449.

Ahrens, C.D. 2008. *Essentials of Meteorology.* Thomson, Belmont, CA.

Alexander, E.B. 1976. Soil temperatures and slope aspect around Hill 998, Tegucigalpa, D.C. *Ceiba.* 20(1): 43–52.

Alexander, E.B. 1994. *Ecological Unit Inventory, Blacks Mountain Experimental Forest.* USDA, Forest Service, Pacific Southwest Research Station, Redding, CA.

Alexander, E.B. 1995. Soil slope-depth relationships in the Klamath Mountains, California and comparisons in central Nevada. *Soil Survey Horizons* 36: 94–103.

Alexander, E.B., R.G. Coleman, T. Keeler-Wolf, and S.P. Harrison. 2007. *Serpentine Geoecology of Western North America.* Oxford University Press, New York.

Alexander, E.B., and R. Poff. 1985. *Soil Disturbance and Compaction in Wildland Soils.* Earth Resources Monograph 8. USDA, Forest Service, Pacific Southwest Region, Vallejo, CA.

Andrews, W.A. (ed.). 1973. *A Guide to the Study of Soil Ecology.* Prentice Hall, Englewood Cliffs, NJ.

Amundson, R. 2001. The carbon budget in soils. *Annual Review of Earth Planetary Science* 29: 535–562.

Arnold, R.W. 1964. Cyclic variations and the pedon. *Soil Science Society of America Proceedings* 28: 801–804.

Atjay, G.L., P. Ketner, and P. Duvigneaud. 1979. Terrestrial primary production and phytomass. In B. Bolin, E.T. Degens, S. Kempe, and P. Ketner (eds.), *The Global Carbon Cycle*, pp. 129–181. Wiley, New York.

Attiwill, P.M., and G.W. Leeper. 1987. *Forest Soils and Nutrient Cycles.* Melbourne University Press, Carlton, Victoria.

Baker, A.J.M., S.P. McGrath, R.D. Reeves, and J.A.C. Smith. 2002. Metal hyper accumulator plants: A review of the ecology and physiology of a biological resource for phytoremediation of metal-polluted soils. In N. Terry and G. Bañuelos (eds.), *Phytoremediation of Contaminated Soil and Water*, pp. 85–107. Lewis Publishers, Boca Raton, FL.

Baker, D.G. 1965. Factors affecting soil temperature. *Minnesota Farm and Home Science* 22(4): 11–13.

Bardgett, R. 2005. *The Biology of Soil*. Oxford University Press, New York.

Barshad, I. 1966. The effect of variation in precipitation on the nature of clay mineral formation in soils from acid and basic igneous rocks. *International Clay Minerals Conference Proceedings* 1: 167–173.

Bates, R.L., and J.A. Jackson (eds.). 1980. *Glossary of Geology*. American Geological Institute, Falls Church, VA.

Bending, G.D., and D.J. Read. 1997. Lignin and soluble phenolic degradation by ericoid and ectomycorrhizal fungi. *Mycological Research* 101: 1348–1354.

Berg, B., and R. Laskowski. 2006. *Litter Decomposition: A Guide to Carbon and Nutrient Turnover*. Elsevier, Amsterdam.

Berner, R.A., R.C. Lasaga, and R.M. Garrels. 1983. The carbonate-silicate geochemical cycle and its effect on atmospheric carbon dioxide over the past 100 million years. *American Journal of Science* 283: 641–683.

Birch, F. 1942. Thermal conductivity and diffusivity. *Geological Society of America Special Paper* 36: 243–266.

Blank, J.R. 1987. *Soil Survey of Wayne County, Indiana*. USDA, Soil Conservation Service (now Natural Resources Conservation Service), Washington, DC.

Bloem, J., D.W. Hopkins, and A. Benedetti (eds.). 2006. *Microbiological Methods for Assessing Soil Quality*. CABI Publishing, Cambridge MA.

Bockheim, J.G., and K.M. Hinkel. 2007. The importance of "deep" organic carbon in permafrost-affected soils of Arctic Alaska. *Soil Science Society of America Journal* 71: 1889–1892.

Bowen, H.J.M. 1979. *Environmental Chemistry of the Elements*. Academic Press, London.

Bowerman, T.S., J. Dorr, S. Leahy, K. Vaga, and J. Warrick. 1997. *Targhee National Forest—Ecological Unit Inventory*. 2 vols. USDA, Forest Service, Targhee National Forest, St. Anthony, ID.

Bowler, J.M. 1973. Clay dunes: their occurrence, formation, and environmental significance. *Earth Sciences Reviews* 9(4): 315–338.

Brewer, R. 1964. *Fabric and Mineral Analysis of Soils*. Wiley, New York.

Bridges, E.M. 1997. Origins, adoption, and development of soil horizon designations. In D.H. Yaalon and S. Berkowicz (eds.), *History of Soil Science*, pp. 47–65. Catena Verlag, Reiskirchen, Germany.

Briggs, D., P. Smithson, K. Addison, and K. Addison. 1997. *Fundamentals of the Physical Environment*. Routledge, London.

Brimblecombe, P. 2004. The global sulfur cycle. In W.H. Schlesinger (ed.), *Biogeochemistry, Treatise on Geochemistry*, pp. 645–682. Vol. 8. Elsevier-Pergamon, Amsterdam.

Brown, A.L. 1978. *Ecology of Soil Organisms*. Heinemann Educational Books, London.

Browne, C.A. 1944. A source book of agricultural chemistry. *Chronica Botanica* 8: 1–290.

Bushnell, T.M. 1943. Some aspects of the soil catena concept. *Soil Science Society of America Proceedings* 7: 466–467.

Busse, M.D., C.J. Shestak, K.R. Hubbert, and E.E. Knapp. 2010. Soil physical properties regulate lethal heating during burning of woody residues. *Soil Science Society of America Journal* 74: 947–955.

Butler, D.R. 1995. *Zoogeomorphology*. Cambridge University Press, Cambridge.

Came, R.E., J.M. Eiler, J. Veizer, K. Azmy, U. Brand, and C.R. Weidman. 2007. Coupling of surface temperatures and atmospheric CO_2 concentrations during the Paleozoic era. *Nature* 449: 198–201.

Canfield, D.E., A.N. Glazer, and P.G. Falkowshi. 2010. The evolution and future of Earth's nitrogen cycle. *Science* 330: 192–196.

Chaney, R.L. 1977. Microelements as related to plant deficiencies and toxicities. In L.F. Elliot and F.J. Stevenson (eds.), *Soils for the Management of Organic Wastes and Waste Waters*, pp. 235–279. American Society of Agronomy, Madison, WI.

Chapin, F.S., III, P. Matson, and H. Mooney. 2002. *Principles of Terrestrial Ecosystem Ecology*. Springer, New York.

Charlson, R.J., T.L. Anderson, and R.E. McDuff. 1992. The sulfur cycle. In S.S. Butcher, R.J. Charlson, G.H. Orians, and G.V. Wolfe (eds.), *Global Biogeochemical Cycles*, pp. 285–300. Academic Press, San Diego.

Chen, C., L. Gomez, and L. Lund. 1991. Acidification potential of snowpack in the Sierra Nevada. *Journal of Environmental Engineering* 117: 472–486.

Christian, C.S. 1957. The concept of land units and land systems. *Ninth Pacific Science Congress Proceedings* 20: 74–81.

Clarke, F.W. 1924. *The Data of Geochemistry*. Bulletin 770. U.S. Geological Survey, Reston, VA.

Coleman, D.C., D.A. Crossley, and P.F. Hendrix. 2004. Fundamentals of Soil Ecology. Elsevier Academic Press, Amsterdam.

Condie, K.C. 1997. *Plate Tectonics and Crustal Evolution*. Butterworth-Heinemann, Oxford.

Dahlgren, R.A. 2007. Acid deposition and acid rain. In W. Chesworth (ed.), *Encyclopedia of Soil Science*, pp. 2–7. Springer, Dordrecht, The Netherlands.

Dahlgren, R.A., J.L. Boettinger, G.L. Huntington, and R.G. Amundson. 1997. Soil development along an elevational transect in the western Sierra Nevada, California. *Geoderma* 78: 207–236.

Darwin, C. 1881. *The Formation of Vegetable Mold through the Action of Worms, with Observations on Their Habitats*. Murray, London.

DeBano, L.F. 2000. The role of fire and soil heating on water repellency in wildland environments: a review. *Journal of Hydrology* 231: 195–206.

Deer, W.A., R.A. Howie, and J. Zussman. 1966. *Introduction to the Rock-Forming Minerals*. Wiley, New York.

Dickens, G.R., M.M. Castillo, and J.C.G. Walker. 1997. A blast of gas in the latest Paleocene: simulating first-order effects of massive dissociation of oceanic methane hydrate. *Geology* 25: 259–262.

Dindal, D.L. (ed.). 1990. *Soil Biology Guide*. Wiley, New York.

Driscoll, C.T., K.M. Driscoll, K.M. Roy, and M.J. Mitchell. 2003. Chemical response of lakes in the Adirondack region of New York to declines in acidic deposition. *Environmental Science and Technology* 37: 2036–2042.

Doran, J.W., and A.J. Jones (eds.). 1996. *Methods of Assessing Soil Quality*. Special Publication 49. Soil Science Society of America, Madison WI.

Eaton, A., and T.R. Beck. 1820. *A Geological Survey of the County of Albany*. Agricultural Society of Albany County, New York.

Edwards, C.A., and P.J. Bohlen. 1996. *Biology and Ecology of Earthworms*. Chapman & Hall, London.

Eskew, D.L., R.M. Welch, and E.E. Cary. 1983. Nickel: an essential micronutrient for legumes and possibly all higher plants. *Science* 222: 621–623.

Eswaran, H., E. Van Den Berg, and P. Reich. 1993. Organic carbon in soils of the world. *Soil Science Society of America Journal* 57: 192–194.

Etherington, J.R. 1982. *Environmental Plant Ecology*. Wiley, London.

Fallou, F.A. 1862. *Pedologie oder allgemeine und besondere Bodenkunde*. G. Schoenfeld, Dresden.

Fanning, D.S., and M.C.B. Fanning. 1989. *Soil Morphology, Genesis, and Classification*. Wiley, New York.

Fettes, D., and J. Desmons (eds.). 2007. *Metamorphic Rocks: A Classification and Glossary of Terms. Recommendations of the International Union of Geological Sciences Subcommission on the Systematics of Metamorphic Rocks*. Cambridge University Press, Cambridge.

Filep, G. 1999. *Soil Chemistry* (English trans.). Akadémiai Kaidó, Budapest.

Finney, H.R., N. Holowaychuk, and M.L Huddleson. 1962. The influence of microclimate on the morphology of certain soils in the Allegheny Plateau of Ohio. *Soil Science Society of America Proceedings* 26: 287–292.

Forman, R.T.T., and M. McGodron. 1986. *Landscape Ecology*. Wiley, New York.

Fox, J.A., and J.L. Hatfield. 1983. *Soil Temperatures in California*. Bulletin 1908. University of California, Division of Agricultural Sciences.

Galloway, J.N. 2004. The global nitrogen cycle. In W.H. Schlesinger (ed.), *Biogeochemistry, Treatise on Geochemistry*, pp. 557–583. Vol. 8. Elsevier-Pergamon, Oxford.

Geiger, R., R.H. Aron, and P. Todhunter. 2003. *The Climate Near the Ground*. Rowan and Littlefield, Lanham, NJ.

Geist, H. 2005. *The Causes and Progression of Desertification*. Ashgate, Burlington, VT.

Glassey, T.F. 1934. Notes regarding soils and vegetation in southwestern Wyoming. *American Soil Survey Association Bulletin* 15: 12–15.

Goranson, R. 1942. Heat capacity; heat of fusion. *Geological Society of America Special Paper* 36: 223–242.

Goudie, A.S. 1988. The geomorphological role of termites and earthworms in the tropics. In H.A. Viles (ed.), *Biogeomorphology*, pp. 166–192. Basil Blackwell, Oxford.

Gruber, N., and J.N. Galloway. 2008. An Earth-system perspective of the global nitrogen cycle. *Nature* 451: 293–296.

Haider, K. 1992. Problems related to the humification process in soils of temperate climates. *Soil Biochemistry* 7: 55–94.

Henderson-Sellers, A., and P.J. Robinson. 1986. *Contemporary Climatology*. Longman, Essex, England.

Hilgard, E.W. 1860. *Report of the Geology and Agriculture of the State of Mississippi*. Jackson, MS.

Hillel, D. 1980. *Fundamentals of Soil Physics*. Academic Press, Orlando, FL.

Hillel, D. 1991. *Out of the Earth: Civilization and Life in the Soil*. Free Press, New York.

Hoffman, K.A. 1988. *American Scientist* 256: 96–103.

Hole, F.D. 1981. Effects of animals on soils. *Geoderma* 25: 75–112.

Howard, R.P., M.J. Singer, and G.A. Frantz. 1981. Effects of soil properties, water content, and compactive effort on the compaction of selected California soils. *Soil Science Society of America Journal* 45: 231–236.

Howells, G. 1995. *Acid Rain and Acid Waters*. Horwood, Hemel Hempstead, Hertfordshire, UK.

Hunter, J.C., and J.E. Horenstein. 1992. The vegetation of the Pine Hill area (California) and its relation to substratum. In A.J.M. Baker, J. Proctor, and R.D. Reeves (eds.), *The Vegetation of Ultramafic (Serpentine) Soils*. Intercept, Andover, Hampshire, England.

Hurt, G.W., and L.M. Vasilas (eds.). 2006. *Field Indicators of Hydric Soils in the United States: A Guide for Indentifying and Delineating Hydric Soils*. Version 6.0. USDA, Natural Resources Conservation Service, Lincoln NE.

Hyndman, D.W. 1985. *Petrology of Igneous and Metamorphic Rocks*. McGraw-Hill Books, New York.

Ismail, F.T. 1970. Biotite weathering and clay formation in arid and humid regions, California. *Soil Science* 109: 257–261.

IUSS Working Group. 2006. *World Reference Base for Soil Resources*. World Soil Resources Report 103. Food and Agricultural Organization of the UN, Rome.

Jackson, K.C. 1970. *Textbook of Lithology*. McGraw-Hill Books, New York.

Jaffe, D.A. 1992. The nitrogen cycle. In S.S. Butcher, R.J. Charlson, G.H. Orians, and G.V. Wolfe (eds.), *Global Geochemical Cycles*, pp. 263–284. Academic Press, London.

Jahnke, R.A. 1992. The phosphorus cycle. In S.S. Butcher, R.J. Charlson, G.H. Orians, and G.V. Wolfe (eds.), *Global Geochemical Cycles*, pp. 301–315. Academic Press, London.

Jenny, H. 1941. *Factors of Soil Formation*. McGraw-Hill, New York.

Jenny, H. 1961. *E. W. Hilgard and the Birth of Modern Soil Science*. Collana Bella Revista Agrochimica, Pisa.

Jenny, H. 1980. *The Soil Resource*. Springer-Verlag, New York.

Jensen, M.E., R.D. Burman, and R.E. Allen (eds.). 1990. *Evapotranspiration and Irrigation Water Requirements*. Manuals and Reports on Engineering Practice 70. American Society of Civil Engineers, New York.

Jenkinson, D.S. 1977. Studies on the decomposition of plant material in soil: The effects of plant cover and soil type on the loss of carbon from ^{14}C labeled rye grass decomposing under field conditions. *Journal of Soil Science* 28: 424–434.

Johnson, D.L. 1990. Biomantle evolution and the redistribution of earth materials and artifacts. *Soil Science* 149: 84–102.

Jones, H.G. 1983. *Plants and Microclimate.* Cambridge University Press, Cambridge.

Jurdant, M., D.S. Lacate, S.C. Zoltai, G.G. Rinka, and R. Wells. 1975. Biophysical land classification in Canada. In B. Bernier and C.H. Winget (eds.), *Forest Soils and Forest Management*, pp. 485–495. Laval University Press, Quebec.

Kabata-Pendias, A. 2011. *Trace Elements in Soils and Plants.* CRC Press, Boca Raton, FL.

Kern, J.S., and M.G. Johnson. 1993. Conservation tillage impacts on national soil and atmospheric carbon levels. *Soil Science Society of America Journal* 57: 200–210.

Killham, K. 1994. *Soil Ecology.* Cambridge University Press, Cambridge.

Kittredge, J. 1948. *Forest Influences.* McGraw-Hill, New York.

Klinka, K., R.N. Green, R.L. Trowbridge, and L.E. Lowe. 1981. *Taxonomic Classification of Taxonomic Forms in Ecosystems of British Columbia.* Land Management Report 8. Ministry of Forests, Land Management, Victoria, BC.

Klyver, F.D. 1931. Major plant communities in a transect of the Sierra Nevada Mountains of California. *Ecology* 12: 1–17.

Krause, H.H., S. Rieger, and S.A. Wilde. 1959. Soils and forest growth on different aspects in the Tanana watershed of interior Alaska. *Ecology* 40: 492–495.

Lal, R. 2009. Sequestering Atmospheric Carbon Dioxide. *Critical Reviews in Plant Sciences* 28(3): 90–96.

Lal, R., T.M. Sobecki, T. Iivari, and J.M. Kimble. 2004. *Soil Degradation in the United States.* Lewis Publishers, Boca Raton FL.

Legrand, M., and R.J. Demas. 1987. A 220-year continuous record of volcanic H_2SO_4 in the Antarctic ice sheet. *Nature* 327: 671–676.

Leifield, J. 2006. Soils as sources and sinks of greenhouse gases. In E. Frossard, W.E.H. Blum, and B. Warkentin (eds.), *Function of Soils for Human Societies and the Environment*, pp. 23–44. Special Publication 266. Geological Society of London.

Leith, H. 1973. Primary production: terrestrial ecosystems. *Human Ecology* 1: 303–332.

Leith, H. 1975. Modeling the primary production of the world. In H. Lieth and R.H. Whittaker (eds.), *Primary Productivity of the Biosphere*, pp. 237–263. Springer-Verlag, New York.

LeMaitre, R.W. (ed.). 1989. *A Classification of Igneous Rocks and Glossary of Terms.* Blackwell Scientific Publications, Oxford.

Leopold, A. 1949. *Sand County Almanac.* Oxford University Press, New York.

Li, Y.-H. 2000. *A Compendium of Geochemistry.* Princeton University Press, Princeton, NJ.

Liebling, R.S., and P.F. Kerr. 1965. Observations on quick clay. *Geological Society of America Bulletin* 76: 953–878.

Likens, G.E., F.H. Borman, R.S. Pierce, J.S. Eaton, and N.M. Johnson. 1977. *Biogeochemistry of a Forested Ecosystem.* Springer-Verlag, Berlin.

Lobry de Bruyn, L.A., and A.J. Conacher. 1990. The role of termites and ants in soil modification: a review. *Australian Journal of Soil Research* 28: 55–93.

Lovelock, J.E. 1993. The soil as a model for the Earth. *Geoderma* 57: 213–215.

Lutz, H.J., and R.F. Chandler. 1946. *Forest Soils.* Wiley, New York.

Ma, Y., and P.S. Hooda. 2010. Chromium, nickel, and cobalt. In P.S. Hooda (ed.), *Trace Elements in Soils,* pp. 461–479. Wiley, New York.

Madigan, M.T., and J.M. Martinko. 2006. *Brock Biology of Microorganisms.* Prentice Hall, Upper Saddle River, NJ.

Margulis, L., and M.J. Chapman. 2009. *Kingdoms and Domains.* Elsevier, Amsterdam.

Marschner, H. 1995. *Mineral Nutrition of Higher Plants.* Academic Press, London.

McKnight, T.L. 1984. *Physical Geography: A Landscape Appreciation.* Prentice-Hall, Englewood Cliffs, NJ.

Meentemeyer, V. 1978. Microclimate and lignin control of decomposition rates. *Ecology* 59: 465–472.

Meyer, J.L. 1992. Seasonal patterns of water quality in blackwater rivers of the Coastal Plain. In C.D. Becker and D.A. Neitzel (eds.), *Water Quality in North American River Systems,* pp. 249–276. Battelle Press, Columbus OH.

Middlemost, E.A.K. 1985. *Magmas and Magmatic Rocks.* Longman, London.

Mitchell, J.K. 1976. *Fundamentals of Soil Behavior.* Wiley, New York.

Mitchell, P.B. 1988. The influences of vegetation, animals, and microorganisms on soil processes. In H.A. Viles (ed.), *Biogeomorphology,* pp. 43–82. Basil Blackwell, Oxford.

Moldenke, A., M. Pajutee, and E. Ingham. 2000. The functional roles of forest soil arthropods: The soil is a living place. In *Forest Soil Biology and Forest Soil Management,* pp. 7–22. Forest Service General Technical Report PSW-GTR-178. USDA, Washington, DC.

Moraghan, J.T., and H.J. Mascagni. 1991. Environmental and soil factors affecting micronutrient deficiencies and toxicities. In J.J. Mortvedt, F.R. Cox, L.M. Shuman, and R.M. Welch (eds.), *Micronutrients in Agriculture,* pp. 371–425. Soil Science Society of America, Madison WI.

Morgan, R.P.C. 2005. *Soil Erosion and Conservation.* Blackwell Science, Oxford.

Nardi, J. 2007. *Life in the Soil.* University of Chicago Press, Chicago.

National Research Council. 1986. *Global Change in the Geosphere-Biosphere.* National Academy Press, Washington, DC.

Nettleton, W.D., K.W. Flach, and G. Borst. 1968. *A Toposequence of Soils in Tonalite Grus in the Southern California Peninsular Range.* Soil Survey Investigations Report 21. USDA, Soil Conservation Service, Washington, DC.

Nicholas, D.J.D. 1975. The functions of trace elements in plants. In D.J.D. Nicholas and A.R. Egan (eds.), *Trace Elements in Soil-Plant-Animal Systems,* pp. 181–198. Academic Press, New York.

Nichols, D.S., and D.H. Boelter. 1984. Fiber size distribution, bulk density, and ash content of peats in Minnesota, Wisconsin, and Michigan. *Soil Science Society of America Journal* 48: 1322–1328.

Norrish, K. 1975. Geochemistry and mineralogy of trace elements. In D.J.D. Nicholas and A.R. Egan (eds.), *Trace Elements in Soil-Plant-Animal Systems,* pp. 56–81. Academic Press, New York.

Ollier, C. 1984. *Weathering.* Longman, Essex, England.

Pais, I., and J.B. Jones Jr. 1997. *The Handbook of Trace Elements*. St. Lucia Press, Boca Raton, FL.

Parrenin, F., V. Masson-Delmonte, P Kröhler, D. Raynaud, D. Paillard, J. Schwander, C. Barbante, A. Landais, A. Wegner, and J. Jouzel. 2013. Synchronous change of atmospheric CO_2 and Antarctic temperature during the last deglacial warming. *Science* 339: 1060–1063.

Paul, E.A., and F.E. Clark. 1996. *Soil Microbiology and Biochemistry*. Academic Press, San Diego.

Pauling, L. 1970. *General Chemistry*. Freeman, San Francisco.

Pearson, W.D. 1992. Historical changes in water quality and fishes of the Ohio River. In C.D. Becker and D.A. Neitzel (eds.), *Water Quality in North American River Systems*, pp. 207–231. Battelle Press, Columbus, OH.

Petit, J.R., J. Jouzel, D. Raynaud, N.I. Barkov, J.M. Barnola, J. Basile, M. Bender, J. Chappellaz, M. Davis, G. Delaygue, M. Delmotte, V.M. Kotlyakov, M. Legrand, L.Y. Lipenkoz, C. Lorius, Pépin, C. Ritz, E. Saltzman, and M. Stievenard. 1999. Climate and atmospheric history of the past 420,000 years from the Vostok ice core, Antarctica. *Nature* 399: 429–435.

Pettijohn, F.J. 1957. *Sedimentary Rocks*. Harper and Row, New York.

Post, W.M., W.R. Emanuel, P.J. Zinke, and A.G. Stangenberger. 1982. Soil carbon pools and world life zones. *Nature* 298: 156–159.

Potts, P.J., A.G. Tindle, and P.C. Webb. 1992. *Geochemical Reference Material Compositions*. CRC Press, Boca Raton, FL.

Qafoku, N.P., VanRanst, E., Noble, A., Baert, G., 2004. Variable charge soils: their mineralogy, chemistry and management. *Advances in Agronomy* 84, 159–215.

Rawls, W.J. 1983. Estimating soil bulk density from particle size analysis and organic matter content. *Soil Science* 135: 123–125.

Reeves, R.D., and R.R. Brooks. 1983. European species of *Thlaspi* L. (Cruciferae) as indicators of nickel and zinc. *Journal of Geochemical Exploration* 18: 275–283.

Renard, K.G., G.R. Foster, G.A. Weesies, and LJ.P. Porter. 1991. RUSLE: Revised Universal Soil Loss Equation. *Journal of Soil and Water Conservation* 46: 30–33.

Retallack, G. 2001. *Soils of the Past*. Blackwell Science, Malden, MA.

Rice, K.C., and J.S. Herman. 2012. Acidification of earth: An assessment across mechanisms and scales. *Applied Geochemistry* 27: 1–14.

Richards, B.N. 1987. *The Microbiology of Terrestrial Ecosystems*. Longman, New York.

Richter, D.D., and D. Markewitz. (1995). How deep is soil? *Bioscience* 45: 600–609.

Rodin, L.E., and N.I Basilevic. 1967. *Production and Mineral Cycling in Terrestrial Vegetation*. Oliver and Boyd, Edinburgh.

Romankevich, E.A. 1990. Biochemical problems of living matter of the present-day biosphere. In V. Ittekkat, S. Kempe, W. Michaelis, and A. Spitey (eds.), *Facets of Modern Biogeochemistry*, pp. 39–51. Springer-Verlag, Berlin.

Rosenberg, N.J. et al. 1983. *Microclimate, the Biological Environment*. Wiley, New York.

Royer, D.L., R.A. Berner, I.P. Montañez, N.J. Tabor, and D.J. Beerling. 2004. CO_2 as a primary driver of Phanerozoic climate. *GSA Today* 14(3): 4–10.

Ruhe, R.V. 1975. *Geomorphology*. Houghton Mifflin, Boston.

Satterlund, D.R., and P.W. Adams. 1992. *Wildland Watershed Management*. Wiley, New York.

Saugier, B., J. Roy, and HA. Mooney. 2001. Estimations of global terrestrial productivity: converging toward a single number? In J. Roy, B. Saugier, and H.A. Mooney (eds.), *Terrestrial Global Productivity*, pp. 543–557. Academic Press, New York.

Scheinost, A.C., and U. Schwertmann. 1999. Color identification of iron oxides and hydroxysufates: Use and limitations. *Soil Science Society of America Journal* 63: 1463–1471.

Schimel, J.P., and J Bennett. 2004. Nitrogen mineralization: challenges of a changing paradigm. *Ecology* 85: 591–602.

Schlesinger, W.H. 1997. *Biogeochemistry: An Analysis of Global Change*. Academic Press, San Diego.

Schlesinger, W.H., J.P. Winkler, and J.P. Megonigal. 2000. Soils and the global carbon cycle. In T.M.L. Wigley and D.S. Schimel (eds.), *The Carbon Cycle*, pp. 93–101. Cambridge University Press, New York.

Schoenmann, L.R. 1923. *Descriptions of Field Methods Followed by the Michigan Land and Economic Survey*. Bulletin IV, vol. 1, pp. 44–52. American Association of Soil Survey Workers, Washington DC.

Schwertmann, U. 1993. Relations between iron oxides, soil color, and soil formation. In J.M. Bigham and E.J. Ciolkosz (eds.), *Soil Color*, pp. 51–69. Special Publication 31. Soil Science Society of America, Madison WI.

Shacklette, H.T., and J.G. Boerngen. 1984. *Element Concentrations in Soils and Other Surficial Materials of the Conterminous United States*. Professional Paper 1270. U.S. Geological Survey, Reston, VA.

Shaler, N.S. 1891. *The Origin and Nature of Soils*. Twelfth Annual Report, Part 1, pp. 219–345. U.S. Geological Survey, Reston, VA.

Shaver, G., and S. Jonasson. 2001. Productivity of Arctic ecosystems. In J. Roy, B. Saugier, and H.A. Mooney (eds.), *Terrestrial Global Productivity*, pp. 204–210. Academic Press, New York.

Shaw, C.F. 1930. Potent factors in soil formation. *Ecology* 11: 239–245.

Shuman, L.M. 1991. Chemical forms of micronutrients in soils. In J.J. Mortvedt, F.R. Cox, L.M. Shuman, and R.M. Welch (eds.), *Micronutrients in Agriculture*, pp. 113–144. Soil Science Society of America, Madison WI.

Sidle, R.C. 1980. *Slope Stability on Forest Land*. Pacific Northwest Extension Publication PNW 209. Oregon State University Extension Service, Corvallis.

Simonson, R.W. 1959. Outline of a theory of soil genesis. *Soil Science Society of America Proceedings* 23: 152–158.

Simpson, M.G. 2007. *Dirt: The Erosion of Civilizations*. University of California Press, Berkeley.

Singer, A. 1989. Palygorskite and sepiolite group minerals. In J.B. Dixon and S.B. Weed (eds.), *Minerals in Soil Environments*, pp. 829–872. Soil Science Society of America, Madison WI.

Smith, G.D., F. Newhall, and L.H. Robinson. 1964. *Soil Temperature Regimes—Their Characteristics and Predictability*. SCS-TP-144. USDA, Soil Conservation Service, Washington, DC.

Soil Survey Staff. 1960. *Soil Classification—A Comprehensive System*. 7th Approximation. USDA, Washington, DC.

Soil Survey Staff. 1975. *Soil Taxonomy: A Basic System for Making and Interpreting Soil Surveys*. Agriculture Handbook 436. USDA, Washington, DC.

Soil Survey Staff. 1993a. *National Soil Survey Handbook*. Title 430-VI. U.S. Government Printing Office, Washington, DC.

Soil Survey Staff. 1993b. *Soil Survey Manual*. USDA Handbook 18. U.S. Government Printing Office, Washington, DC.

Soil Survey Staff. 1999. *Soil Taxonomy: A Basic System for Making and Interpreting Soil Surveys*. Agriculture Handbook 436. USDA, Washington, DC.

Soil Survey Staff. 2010. *Keys to Soil Taxonomy*. USDA, Washington, DC.

Sorensen, S.L.P. 1909. Über die Messung und die Bedeutung der Wasserstoffkonzentration bei enzymatischen Prozessen. *Biochem Zeit* 21: 131–139.

Speidel, D.H., and A.F. Agnew. 1982. *The Natural Geochemistry of Our Environment*. Westview Press, Boulder, CO.

Stevenson, F.J. 1994. *Humus Chemistry, Genesis, Compositions, Reactions*. Wiley, New York.

Stevenson, F.J., and M.A. Cole. 1999. *Cycles of Soil: Carbon, Nitrogen, Phosphorous, and Micronutrients*. Wiley, New York.

Stoops, G. 2003. *Guidelines for Analysis and Descriptions of Soil and Regolith Thin Sections*. Soil Science Society of America, Madison, WI.

Strahler, A.N., and A.H. Stahler. 1987. *Modern Physical Geography*. Wiley, New York.

Streckeisen, A.L. 1979. Classification and nomenclature of volcanic rocks, lamprophyres, carbonatites, and melilitic rocks: Recommendations and suggestions of the IUGS Subcommission on the Systematics of Igneous Rocks. *Geology* 7: 331–335.

Strom, R. 2007. *Hot House: Global Climate Change and the Human Condition*. Copernicus Books, Springer, New York.

Sundquist, E.T., K.V. Ackerman, and L. Parker. 2009. An introduction to global carbon cycle management. In B.J. McPherson and E.T. Sundquist (eds.), *Carbon Sequestration and Its Role in the Global Carbon Cycle*, pp. 1–23. American Geophysical Union, Washington, DC.

Sundquist, E.T., and R.F. Keeling. 2009. The Mauna Loa carbon dioxide record: lessons for long-term earth observations. In B.J. McPherson and E.T. Sundquist (eds.), *Carbon Sequestration and Its Role in the Global Carbon Cycle*, pp. 27–35. American Geophysical Union, Washington, DC.

Svalberg, T., D. Tart, and D. Fallon. 1997. *Bridger-East Ecological Unit Inventory*. 3 vols. USDA, Forest Service, Bridger-Teton National Forest, Pinedale, WY.

Swanson, F.J., T.K. Kratz, N. Caine, and R.G. Woodmansee. 1988. Landform effects on ecosystem patterns and processes. *Bioscience* 38: 92–98.

Swift, M.J., O.W. Heal, and J.M. Anderson. 1979. *Decomposition in Terrestrial Ecosystems*. University of California Press, Berkeley.

Tandarich, J.P., R.G. Darmody, and L.R. Folmer. 1988. The development of pedologic thought: some people involved. *Physical Geography* 9: 162–174.

Taylor, S.R., and S.M. McLennan. 1985. *The Continental Crust: Its Composition and Evolution*. Blackwell Science, Boston.

Thornthwaite, C.W. 1948. An approach toward a rational classification of climate. *Geographical Review* 38: 55–94.

Thornthwaite, C.W., and J.R. Mather. 1955. The water balance. *Climatology* 8(1): 174.

Thorp, J. 1942. The influence of environment on soil formation. *Soil Science Society of America Proceedings* 6: 39–46.

Toy, T.J., and R.F. Hadley. 1987. *Geomorpholog and Reclamation of Disturbed Lands*. Academic Press, New York.

Travis, R.B. 1955. Classification of rocks. *Colorado School of Mines Quarterly* 50(1): 1–98.

Trewartha, G.T. 1968. *An Introduction to Climate*. McGraw-Hill, New York.

Troll, C. 1971. Landscape ecology (geoecology) and biogeocoenology—a terminological study. *Geoforum* 8: 43–46.

Tucker, M.E. 1991. *Sedimentary Petrology*. Blackwell Scientific Publishers, Oxford.

Tudge, C. 2000. *The Variety of Life*. Oxford, New York.

Van Breeman, N. 1993. Soils as biotic constructs favoring net primary productivity. *Geoderma* 57: 183–211.

Van Breeman, N., and P. Buurman. 1998. *Soil Formation*. Kluwer, Dordrecht, The Netherlands.

Van Oost, K., H. Van Hemelryck, and J.W. Harden. 2009. In B.J. McPherson and E.T. Sundquist (eds.), *Carbon Sequestration and Its Role in the Global Carbon Cycle*, pp. 189–202. American Geophysical Union, Washington, DC.

Veatch, J.O. 1937. The idea of the natural land type. *Soil Science Society of America Proceedings* 2: 499–503.

Vinogradov, A.P. 1962. Average contents of chemical elements in the principle types of igneous rocks of the Earth's crust. *Geochemistry* (English translation), 1962(7): 641–664.

Vitousek, P.M., O.A. Chadwick, T.E. Crews, J.H. Fownes, D.M. Hendricks, and D.Herbert. (1997). Soil and ecosystem development across the Hawaiian Islands. *GSA Today* 7(9): 1–8.

Wahrhaftig, C. 1965. Stepped topography of the southern Sierra Nevada, California. *Geological Society of America Bulletin* 76: 1165–1190.

Waksman, S.A. 1932. *Humus*. Williams and Wilkins, Baltimore.

Walker, T.W., and S.K. Syers. 1976. The fate of phosphorous during pedogenesis. *Geoderma* 15: 1–19.

Walter, H. 1974. *Vegetation of the Earth in Relation to Climate and the Ecophysiological Conditions*. Springer, New York.

Welch, R.M., W.H. Allaway, W.A. House, and J. Kubota. 1991. In J.J. Mortvedt, F.R. Cox, L.M. Shuman, and R.M. Welch (eds.), *Micronutrients in Agriculture*, pp. 31–57. Soil Science Society of America, Madison, WI.

Wellburn, A. 1994. *Air Pollution and Climate: The Biological Impact*. Longman, New York.

Wertz, W.A., and J.F. Arnold. 1972. *Land Systems Inventory*. USDA, Forest Service, Intermountain Region, Ogden, UT.

Whittaker, R.H. 1959. On the broad classification of organisms. *Quarterly Review of Biology* 34: 210–226.

Whittaker, R.H. 1975. *Communities and Ecosystems*. Macmillan, New York.

Williams, H., J. Turner, and C.M. Gilbert. 1982. *Petrography*. Freeman, San Francisco. Williams, R.J.P., and J.J.R. Fraústo da Silva. 1996. *The Natural Selection of the Chemical Elements*. Oxford University Press, Oxford.

Winiwarter, V. 2006. Prolegomena to a History of Soil Knowledge in Europe. In J.R. McNeil and V. Winiwarter (eds.), *Soils and Society*, pp. 177–215. White Horse Press, Isle of Harris, UK.

Wischmeier, W.H., and D.D. Smith. 1978. *Predicting Rainfall Erosion Losses*. Agricultural Handbook 537. USDA, Washington, DC.

Woese, C., and G. Fox G. 1977. Phylogenetic structure of the prokaryotic domain: the primary kingdoms. *Proceedings of the National Academy of Sciences, USA*, 74(11): 5088–5090.

Wood, M. 1995. *Environmental Soil Biology*. Blackie Academic and Professional, London.

Zhu, X.G., S.P. Long, and D.R. Ort. 2008. What is the maximum efficiency with which photosynthesis can convert solar energy into biomass? *Current Opinion in Biotechnology* 19: 153–159.

Index